SpringerBriefs in Mathematical Physics

Volume 37

SpringerBriefs are characterized in general by their size (50–125 pages) and fast production time (2–3 months compared to 6 months for a monograph).
Briefs are available in print but are intended as a primarily electronic publication to be included in Springer's e-book package.

Typical works might include:

- An extended survey of a field
- A link between new research papers published in journal articles
- A presentation of core concepts that doctoral students must understand in order to make independent contributions
- Lecture notes making a specialist topic accessible for non-specialist readers.

SpringerBriefs in Mathematical Physics showcase, in a compact format, topics of current relevance in the field of mathematical physics. Published titles will encompass all areas of theoretical and mathematical physics. This series is intended for mathematicians, physicists, and other scientists, as well as doctoral students in related areas.

Editorial Board

- Nathanaël Berestycki (University of Cambridge, UK)
- Mihalis Dafermos (University of Cambridge, UK / Princeton University, US)
- Atsuo Kuniba (University of Tokyo, Japan)
- Matilde Marcolli (CALTECH, US)
- Bruno Nachtergaele (UC Davis, US)
- Hirosi Ooguri (California Institute of Technology, US / Kavli IPMU, Japan)

Springer Briefs in a nutshell

SpringerBriefs specifications vary depending on the title. In general, each Brief will have:

- 50–125 published pages, including all tables, figures, and references
- Softcover binding
- Copyright to remain in author's name
- Versions in print, eBook, and MyCopy

More information about this series at http://www.springer.com/series/11953

Hitoshi Konno

Elliptic Quantum Groups

Representations and Related Geometry

Springer

Hitoshi Konno
Department of Mathematics
Tokyo University of Marine Science
and Technology
Tokyo, Japan

ISSN 2197-1757 ISSN 2197-1765 (electronic)
SpringerBriefs in Mathematical Physics
ISBN 978-981-15-7386-6 ISBN 978-981-15-7387-3 (eBook)
https://doi.org/10.1007/978-981-15-7387-3

This Springer imprint is published by the registered company Springer Nature Singapore Pte Ltd.
The registered company address is: 152 Beach Road, #21-01/04 Gateway East, Singapore 189721, Singapore

There must be a way only visible to those who reach there.

Preface

Quantum group is an algebraic system associated with a solution of the Yang–Baxter equation (YBE). In general, it is an associative algebra classified by a finite dimensional simple Lie algebra $\bar{\mathfrak{g}}$ or an affine Lie algebra $\mathfrak{g}$, or even a toroidal algebra $\mathfrak{g}_{tor}$, and equipped with a certain co-algebra structure. Typical examples are Yangians $Y(\bar{\mathfrak{g}})$ and affine quantum group $U_q(\mathfrak{g})$, which are quantum groups associated with a rational and a trigonometric solution of the YBE, respectively. See, for example, [21, 35, 122].

Similarly, an elliptic quantum group is a quantum group associated with an elliptic solution of the YBE and classified by an affine Lie algebra $\mathfrak{g}$. In elliptic setting, there are two types of YBEs: the vertex type and the face type (Figs. 1 and 2). These are originated in the two types of solvable lattice models: the vertex model and the face model.[1] They are statistical mechanical models on a two-dimensional square lattice defined by assigning Boltzmann weights to each vertex and face, respectively. Typical examples are the eight-vertex model [12] and the eight-vertex SOS model [7], respectively. Accordingly, there are two types of elliptic quantum groups: the vertex type and the face type [77]. In particular, since the face type YBE is equivalent to the so-called dynamical YBE (DYBE) [45], the face type elliptic quantum group provides an important example of dynamical quantum groups.[2] See Sects. 1.2.3 and 1.2.4.

The aim of this book is to give a survey of the recent developments on elliptic quantum groups, their representations and related geometry. As the simplest example, we mainly consider the face type elliptic quantum group $U_{q,p}(\widehat{\mathfrak{sl}}_2)$ associated with an affine Lie algebra $\widehat{\mathfrak{sl}}_2 = \widehat{\mathfrak{sl}}(2, \mathbb{C})$. It is an elliptic and dynamical analogue of the affine quantum group $U_q(\widehat{\mathfrak{sl}}_2)$ in the Drinfeld realization [30]. The subject includes a brief history of formulations and applications, a detailed

[1]The face model is sometimes called the solid-on-solid (SOS) model or the interaction-round-a-face (IRF) model.

[2]Strictly speaking, the vertex type elliptic quantum groups are also dynamical, where the elliptic nome p gets a shift by a central element q^{2c}. See Sect. 1.2.4.

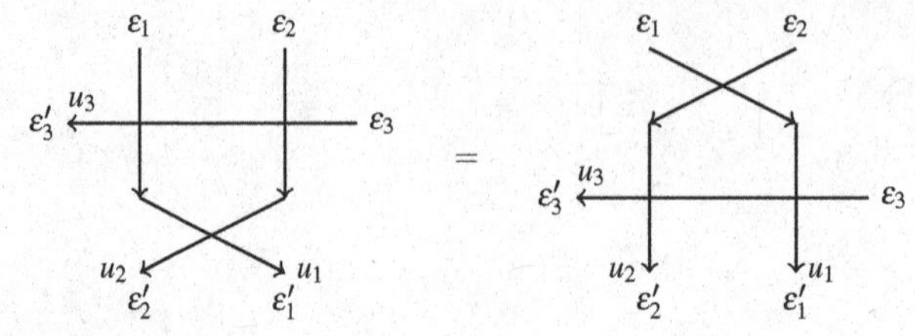

Fig. 1 The vertex type Yang–Baxter equation

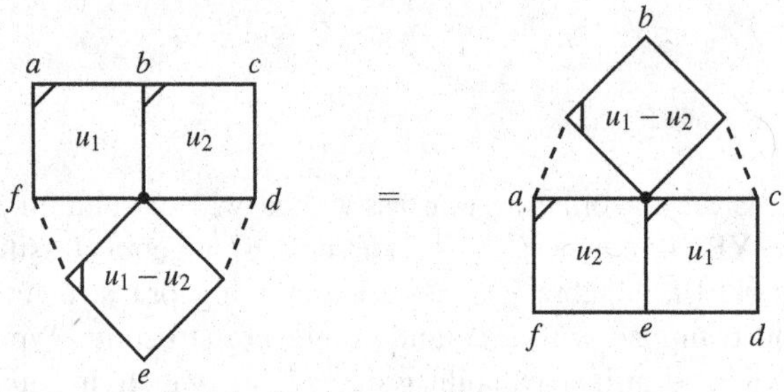

Fig. 2 The face type Yang–Baxter equation

formulation of the elliptic quantum group $U_{q,p}(\widehat{\mathfrak{sl}}_2)$, an explicit construction of both finite and infinite dimensional representations, and of the vertex operators, a derivation of the elliptic q-Knizhnik–Zamolodchikov (KZ) equations and their elliptic hypergeometric integral solutions, as well as their geometric interpretation.

In particular, the vertex operators allow us to derive the so-called weight functions. The weight functions in general appear in hypergeometric integral solutions to the (q-)KZ equations [50, 115, 119, 120, 153] and form a basis of the (q-)twisted de Rahm cohomology of the integrals. See, for example, [5]. Recently, it has been shown that the weight functions can be identified with Okounkov's stable envelopes for equivariant cohomology $\mathrm{H}_T^*(X)$, K-theory $\mathrm{K}_T(X)$, and elliptic cohomology $\mathrm{E}_T(X)$ of a corresponding quiver variety X [51, 69, 100, 135–137]. The latter objects provide a basis of these equivariant cohomologies[3] and play an important role to construct geometric representations of quantum groups. In addition, they provide a new geometric idea of dealing with quantum integrable systems in connection with the Alday–Gaiotto–Tachikawa correspondence between 4D SUSY gauge theories and the CFTs [2], the Nekrasov–Shatashvili correspondences between quantum integrable systems and quantum cohomology [129] as well as the mirror duality of 3D SUSY gauge theories [3, 134]. They are rapidly growing topics in modern

[3] At least for $\mathrm{H}_T^*(X)$ and $\mathrm{K}_T(X)$, stable envelopes provide a good basis of the corresponding algebra [117, 130]. However for $\mathrm{E}_T(X)$ [3], it involves more discussion.

Mathematical Physics. To guide the reader to the entrance of such topics is also an aim of this book.

This book is organized as follows:

Chapter 1 serves as an introduction to "the elliptic problems" related to elliptic quantum groups. It includes quantum integrable systems, SUSY gauge theory, (quantum) equivariant cohomology, and deformed W-algebras. A brief history of elliptic quantum groups is also given. There are some different formulations. We classify them by their generators and co-algebra structures : the Quasi-Hopf algebra formulation $\mathscr{A}_{q,p}(\widehat{\mathfrak{sl}}_N)$ (the vertex type), $\mathscr{B}_{q,\lambda}(\mathfrak{g})$ (the face type), the Faddeev–Reshetikhin–Sklyanin–Semenov–Tian–Shansky–Takhtajan (FRST) formulation $E_{q,p}(\widehat{\mathfrak{sl}}_N)$ (the face type), and the Drinfeld realization $U_{q,p}(\mathfrak{g})$ (the face type).

In Chap. 2, the elliptic quantum group $U_{q,p}(\widehat{\mathfrak{sl}}_2)$ is defined by generators and relations. We also construct the L-operators satisfying the dynamical RLL-relation with respect to the elliptic dynamical R-matrix and show a consistency to the quasi-Hopf formulation $\mathscr{B}_{q,\lambda}(\widehat{\mathfrak{sl}}_2)$. The isomorphism between $U_{q,p}(\widehat{\mathfrak{gl}}_2)$ and the central extension $E_{q,p}(\widehat{\mathfrak{gl}}_2)$ of Felder's elliptic quantum group is discussed briefly in Appendix C.

In Chap. 3, we describe a co-algebra structure of $U_{q,p}(\widehat{\mathfrak{sl}}_2)$ given as an H-Hopf algebroid. This provides a convenient framework to deal with the dynamical shift in various representations of $U_{q,p}(\widehat{\mathfrak{sl}}_2)$. See Chaps. 4–8.

In Chap. 4, we introduce the notion of dynamical representation as representation of $U_{q,p}(\widehat{\mathfrak{sl}}_2)$. As examples, the evaluation representation associated with the vector representation and the level-1 highest weight representations are presented.

In Chap. 5, we introduce the vertex operators as intertwining operators of $U_{q,p}(\widehat{\mathfrak{sl}}_2)$-modules. Applying the representations constructed in Chap. 4, we obtain a realization of the vertex operators explicitly. Exchange relations among the vertex operators are also given.

Chapter 6 is devoted to a discussion of the elliptic weight functions. Their derivation is given by considering a composition of the vertex operators and applying their explicit realizations obtained in Chap. 5. We then discuss their basic properties such as triangular property, transition property, orthogonality, quasi-periodicity, and the shuffle product structure.

In Chap. 7, we discuss tensor product representations of the evaluation representation constructed in Chap. 4. Introducing the so-called Gelfand–Tsetlin basis, we construct an action of $U_{q,p}(\widehat{\mathfrak{sl}}_2)$ on it. Remarkably, the change of basis matrix from the standard basis to the Gelfand–Tsetlin basis is given by a specialization of the elliptic weight functions.

In Chap. 8, we show that a trace of a composition of the vertex operators over the level-1 highest weight representations satisfies the face type elliptic q-KZ equation. By calculating the trace explicitly, we obtain an elliptic hypergeometric integral solution of it.

In Chap. 9, we discuss a geometric interpretation of the results in Chaps. 6–8. For the equivariant elliptic cohomology $\mathrm{E}_T(X_k)$ of the cotangent bundle of the Grassmannian variety $X_k = T^*\mathrm{Gr}(k,n)$, we identify the elliptic weight functions with the elliptic stable envelopes and find a remarkable correspondence between the

Gelfand-Tsetlin basis and the standard basis in Chap. 7 to the fixed point classes and the stable classes, respectively. This allows us to interpret the tensor product representation in Chap. 7 as an action of $U_{q,p}(\widehat{\mathfrak{sl}}_2)$ on $\bigoplus_k \mathrm{E}_T(X_k)$. There are five appendices: Appendix A is a summary of the Drinfeld–Jimbo formulation of the affine quantum group $U_q(\mathfrak{g})$ and its universal R-matrix. In Appendix B, we present a definition of the elliptic quantum group $U_{q,p}(\widehat{\mathfrak{gl}}_2)$. In Appendix C, we give the central extension $E_{q,p}(\widehat{\mathfrak{gl}}_2)$ of Felder's elliptic quantum group and discuss the isomorphism between $U_{q,p}(\widehat{\mathfrak{gl}}_2)$ and $E_{q,p}(\widehat{\mathfrak{gl}}_2)$ briefly. Appendix D is devoted to a proof of Theorem 7.2.3. Appendix E is a collection of formulas for calculating a trace of operators on the Fock space.

Acknowledgements

This book is an outgrowth of a series of lectures delivered at Hiroshima University, June 1998, Kobe University, December 2014, the University of Tokyo, October 2019, as well as several seminars at RIMS, Kyoto University, and Perimeter Institute for Theoretical Physics. I would like to thank Tomoyuki Arakawa, Giovanni Felder, Omar Foda, Vassily Gorbounov, Tatsuyuki Hikita, Masahiko Ito, Michio Jimbo, Syu Kato, Taro Kimura, Yoshiyuki Kimura, Anatol Kirillov, Takeo Kojima, Christian Korff, Atsuo Kuniba, Michael Lashkevich, Atsushi Matsuo, Leonardo Mihalcea, Tetsuji Miwa, Alexander Molev, Kohei Motegi, Hiraku Nakajima, Atsushi Nakayashiki, Hiroshi Naruse, Andrei Negut, Masatoshi Noumi, Masato Okado, Tadashi Okazaki, Andrei Okounkov, Kazuyuki Oshima, Yaroslav Pugai, Hjalmar Rosengren, Yoshihisa Saito, Junichi Shiraishi, Andrey Smirnov, Vitaly Tarasov, Oleksandr Tsymbaliuk, Alexander Varchenko, Robert Weston, and Yasuhiko Yamada for valuable discussions.

Tokyo, Japan Hitoshi Konno
January 2020

Contents

Chapter 1
Introduction

This chapter provides a brief introduction to the subjects related to elliptic quantum groups including quantum integrable systems, SUSY gauge theory, (quantum) equivariant cohomology, and deformed W-algebras. A brief history of elliptic quantum groups is also given. There are some different formulations developed independently and sometimes dependently. They are classified by their generators and co-algebra structures into the following three : the Quasi-Hopf-algebra formulation $\mathscr{A}_{q,p}(\widehat{\mathfrak{sl}}_N)$ (the vertex type), $\mathscr{B}_{q,\lambda}(\mathfrak{g})$ (the face type), the FRST formulation $E_{q,p}(\widehat{\mathfrak{sl}}_N)$ (the face type), and the Drinfeld realization $U_{q,p}(\mathfrak{g})$ (the face type).

1.1 Why Elliptic?

There are several reasons which tempt us to study elliptic quantum groups. We here list some of them.

1.1.1 The Top of the Hierarchy

It is well known that there are the rational, trigonometric, and elliptic solutions of the YBE, i.e. the R-matrices given in terms of rational, trigonometric, and elliptic functions, respectively. These R-matrices are used to formulate 2-dimensional solvable lattice models or related 1-dimensional quantum spin chain models such as the XXX, XXZ, and XYZ models. An algebraic structure associated with these R-matrices is nothing but the quantum group.

H. Konno, *Elliptic Quantum Groups*, SpringerBriefs in Mathematical Physics 37,
https://doi.org/10.1007/978-981-15-7387-3_1

It is remarkable that such a rational–trigonometric–elliptic hierarchy exists in various subjects including

- quantum many body integrable systems such as Calogero–Sutherland–Moser differential models as well as Ruijsenaars difference systems,
- ordinally, q-, and elliptic hypergeometric series and integrals,
- (equivariant) cohomology $H_T^*(X)$, K-theory $\mathrm{K}_T(X)$, and elliptic cohomology $\mathrm{E}_T(X)$ and related geometric representation theories of quantum groups on $H_T^*(X)$ and $\mathrm{K}_T(X)$ [64, 126, 158] and a conjecture in the elliptic case [65, 70].
- SUSY gauge theories with the Ω-background by Nekrasov–Shatashvili in 4D, 5D and 6D.

In all cases the elliptic one is on the top of the hierarchy and the other cases are obtained formally as a degeneration limit of it. It is very fascinating to remark that these subjects seem to be related to each other deeply, known partially as the Alday–Gaiotto–Tachikawa correspondence, the Nekrasov–Shatashvili correspondences and so on [2, 4, 66, 67, 117, 124, 129].

1.1.2 Natural Appearance of the Dynamical Parameters

The dynamical parameters first appeared in the face type elliptic solvable lattice models and so as in the face type elliptic quantum groups [45, 76, 77]. However since then it has been a long-standing problem to find the role of them in further subjects. Recently there are two important developments.

The first one is in a formulation of quantum equivariant cohomology theory. In [117], Maulik–Okounkov formulated a quantum equivariant cohomology by identifying the quantum multiplication with a quantum connection appearing in certain quantum difference equation, which is compatible with the q-KZ equation. Such quantum difference equation turns out a difference equation in the dynamical parameters, whereas the q-KZ equation is the one in the equivariant parameters. Hence the dynamical parameters were identified with the quantum parameters in the quantum equivariant cohomology.

The second development was led by Aganagic–Okounkov [3]. They identified the dynamical parameters with the Kähler parameters in a product of elliptic curves $\mathrm{Pic}(X) \otimes_{\mathbb{Z}} E$ in their extended equivariant elliptic cohomology $\mathrm{E}_T(X)$ of quiver variety X. They also found that in the Mirror duality of 3D SUSY gauge theories the roles of the equivariant and the Kähler parameters in $\mathrm{E}_T(X)$ are exchanged as the mass and the Fayet–Iliopoulos parameters, respectively. One should note that this phenomenon is an example of the deeper subject, the symplectic duality.

Thanks to these, it turns out that the dynamical parameters play important roles in various subjects in Mathematical Physics in particular listed above. The face type elliptic quantum group should be useful for further developments in these subjects as a natural algebraic framework of dealing with both the equivariant[1] and the dynamical parameters.

1.1.3 Connection to Deformed W-Algebras

It is well known that the critical behavior of the face type elliptic solvable lattice models associated with the affine Lie algebra $\mathfrak{g}$ [25, 26, 81, 106, 107] is described by the W algebra $W(\bar{\mathfrak{g}})$ of the coset type $\mathfrak{g}_{r-g-k} \oplus \mathfrak{g}_k \supset \mathfrak{g}_{r-g}$ [18, 68, 112]. Here r is a real positive parameter called the restriction height, k in $\mathfrak{g}_k$ denotes the level of the representation, and g denotes the dual Coxeter number. Accordingly the face type elliptic quantum group $U_{q,p}(\mathfrak{g})$ has a deep connection to deformation of the W-algebra $W_{p,p^*}(\bar{\mathfrak{g}})$ of the same coset type, where $p = q^{2r}$ and $p^* = q^{2(r-k)}$ play the same role as the standard parameters q and t, respectively [9, 41, 57].

In particular, at least the level-1 $U_{q,p}(\mathfrak{g})$ contains the algebras of the screening currents of $W_{p,p^*}(\bar{\mathfrak{g}})$ as the two nilpotent parts [41, 76, 94]. The reason why such peculiar connection happens is that $U_{q,p}(\mathfrak{g})$ is indeed a q-deformation of the Feigin–Fuchs construction of the Virasoro or the W-algebras of the coset type in CFT [11, 27, 42, 68, 84, 132]. The Feigin–Fuchs construction is a method of obtaining screening currents of the W algebra $W(\bar{\mathfrak{g}})$ from the currents of the corresponding untwisted affine Lie algebra $\mathfrak{g}$ in the Lepowsky–Wilson realization [110].[2] It deforms the free bosons appearing in the currents of $\mathfrak{g}$ by introducing the background charges and changes them to the so-called Feigin–Fuchs bosons. The background charges depend on a parameter r,[3] and the elliptic nome $p = q^{2r}$ in $U_{q,p}(\mathfrak{g})$ is nothing but a q-deformation of this parameter. In addition, in this process the Z-algebra part is unchanged except for the zero-mode parts of the free bosons, which is modified by the background charges. This modification makes the Z-algebra dynamical. Similarly, $U_{q,p}(\mathfrak{g})$ has the deformed Z-algebra which is the same as the one of the quantum affine algebra $U_q(\mathfrak{g})$ [17, 55, 82, 116] except for it becomes dynamical [38, 76, 94].

---

[1] In representation theory of the face type elliptic quantum group, the equivariant parameters are identified with evaluation parameters in the level-0 representation. See Chap. 9.

[2] There the currents of the level-k representation of $\mathfrak{g}$ are realized in terms of rank $\bar{\mathfrak{g}}$ free bosons and the level-k Z-algebra. Note that in the Lepowsky–Wilson realization free boson means only its non-zero modes part. Then the Z-algebra can be regarded as a tensor product of the so-called parafermion algebra [160] and the zero-mode parts of the free bosons.

[3] An effect of this deformation can be seen in a change of the central charge. For example, $\mathfrak{g} = \widehat{\mathfrak{sl}}_2$ of level k with the central charge $c_k = \frac{3k}{k+2}$ is deformed to the coset Virasoro model $(\widehat{\mathfrak{sl}}_2)_{r-2-k} \oplus (\widehat{\mathfrak{sl}}_2)_k \supset (\widehat{\mathfrak{sl}}_2)_{r-2}$ with the central charge $c_{Vir} = \frac{3k}{k+2}\left(1 - \frac{2(k+2)}{r(r-k)}\right)$.

1.1.4 Rigorous Formulation of the Massless Regime of Spin Chain Models and 2D Integrable Quantum Field Theories

One of the most successful application of the representation theory of quantum groups is the algebraic analysis of solvable lattice models developed by Jimbo and Miwa and their collaborators [79]. For example, for the anti-ferromagnetic regime of the XXZ spin chain, the spaces of physical states satisfying certain boundary conditions are identified with the level-1 highest weight representations of $U_q(\widehat{\mathfrak{sl}}_2)$. See Sect. 1.2.2 for more details. This regime is specified by the anisotropic parameter $\Delta = \frac{q+q^{-1}}{2} < -1$, and is also known to be obtained from the XYZ spin chain in the principal regime by taking the trigonometric limit $p \to 0$.

A similar formulation of the face type elliptic lattice models has been done by using representation theory of $U_{q,p}(\mathfrak{g})$. See for example [19, 20, 93] and references therein.

Moreover there are some other important models obtained from the same XYZ spin chain but in the different regime, the disordered regime, and by taking the other trigonometric limit $p \to 1$, i.e. the degeneration limit of the conjugate modulus. Those are the sine-Gordon model, which is known as a typical example of 2D integrable quantum field theory, and the XXZ spin chain in the massless regime, which is specified by $|\Delta| \leq 1$ [12, 20]. One naive guess was that representations of $U_q(\widehat{\mathfrak{sl}}_2)$ with $|q| = 1$ could be applied to a direct formulation of these models. However it turns out that such representations have no discrete grading and behave badly, so one needs some regularizations [75, 111].

One can avoid such difficulties and keep everything rigorous, if one starts from the elliptic setting, i.e. the XYZ spin chain. The basic strategy, for example used in [19, 20], is that one formulates all physical quantities, such as correlation functions and form factors, of the XYZ spin chain in the principal regime, where both q and p are generic, by using representations of $U_{q,p}(\widehat{\mathfrak{sl}}_2)$ and the vertex-face correspondence [7, 12, 109], and then maps them to those of the XYZ disordered regime and takes their trigonometric limit $p \to 1$ with scaling the spectral parameters. In [20], the difference between the two scaling limits, which lead to the sine-Gordon and the massless XXZ spin chain, respectively, is also explained.

1.2 Brief History of Elliptic Quantum Groups

We next try to describe a history of elliptic quantum groups. There are some different formulations of them. We make a brief summary of them with some historic comments. The following is not a complete description at all.

1.2.1 Sklyanin Algebra

A research on elliptic quantum algebra was initiated by Sklyanin [144, 145]. He considered an algebra $\mathscr{A}$ associated with Baxter's elliptic R-matrix $R(u) \in \mathrm{End}(\mathbb{C}^2 \otimes \mathbb{C}^2)$ [12].

Let $L(u) \in \mathrm{End}(\mathbb{C}^2) \otimes \mathscr{A}$ and assume that $L(u)$ takes the form

$$L(u) = \sum_{a=0,1,2,3} w_a(u)\sigma^a \otimes S^a,$$

where σ^a denote the Pauli matrices with $\sigma^0 = I$ and $w_a(u)$'s are certain functions of u determined by requiring that the vector representation $(\mathrm{id} \otimes \pi_V)L(u) = \sum_a w_a(u)\sigma^a \otimes \sigma^a$ coincides with $R(u)$. Then he found that the RLL-relation

$$R^{(12)}(u_1 - u_2)L^{(1)}(u_1)L^{(2)}(u_2) = L^{(2)}(u_2)L^{(1)}(u_1)R^{(12)}(u_1 - u_2), \quad (1.2.1)$$

$$L^{(1)}(u) = L(u) \otimes \mathrm{id}, \qquad L^{(2)}(u) = \mathrm{id} \otimes L(u) \quad (1.2.2)$$

yields the relations among S^a's, which are *independent* from the spectrum parameters u_i $(i = 1, 2)$. The algebra $\mathscr{A}$ generated by S^a's subject to these relations are called the Sklyanin algebra. His algebra has a lot of interesting properties and applications. See [138, 151] for recent progress. However its co-algebra structure is still missing.

1.2.2 The Algebraic Analysis and Elliptic Algebra $\mathscr{A}_{q,p}(\widehat{\mathfrak{sl}}_N)$ and $U_{q,p}(\mathfrak{g})$

In early 1990s, Jimbo, Miwa, and their collaborators developed the algebraic analysis, which is a new method of solving exactly solvable lattice models by using representation theory of affine quantum groups $U_q(\mathfrak{g})$ [79].

The essence of this method is to consider the model on the infinite lattice directly and give it a "regularized" picture, i.e. a mathematically well-defined formulation, by using both finite and infinite dimensional representations of $U_q(\mathfrak{g})$. In particular, infinite dimensional representations are identified with the space of states of the model specified by the ground state boundary condition, where the configuration at enough far sites from the center of the lattice is given by the ground state configuration. In addition the two types intertwining operators (vertex operators) of $U_q(\mathfrak{g})$-modules realize the local operators of the model, such as spin operators, on the infinite dimensional representations (type I) and the creation operators of the physical excitations (type II), respectively. Then this formulation allows us to calculate correlation functions as well as form factors of the models. It is also remarkable that these physical quantities are characterized as solutions of q-KZ equations [35, 56, 72]. Hence the algebraic analysis can be regarded as an off-critical extension of the celebrated 2D conformal field theory (CFT) initiated by Belavin, Polyakov, and Zamolodchikov [15].

Then it was the next step to formulate the elliptic lattice models both of the vertex type and the face type in the same spirit. Note that the elliptic vertex models exist only for an A-type affine Lie algebra [13, 14], whereas the face models exist for any affine Lie algebras [81, 95, 106, 107].

There were in fact some attempts to formulate elliptic lattice models by using a graphical argument based on Baxter's corner transfer matrix and the lattice vertex operators ("the half transfer matrices")[54, 80]. One of the important results in them is a derivation of the difference equations for correlation functions, which are now known as the elliptic q-KZ equations both of the vertex type and the face type.

Motivated by this, the first attempt to formulate an elliptic quantum group was done for the vertex type in 1994 [52, 53]. There the elliptic algebra $\mathscr{A}_{q,p}(\widehat{\mathfrak{sl}}_2)$ was proposed in the FRST formulation. Roughly speaking $\mathscr{A}_{q,p}(\widehat{\mathfrak{sl}}_2)$ is an associative algebra generated by the coefficients of the L-operator $L^+(\zeta)$ in formal Laurent series in ζ subject to the RLL-relation

$$R^{+(12)}(\zeta_1/\zeta_2, p^{1/2})L^{+(1)}(\zeta_1)L^{+(2)}(\zeta_2) = L^{+(2)}(\zeta_2)L^{+(1)}(\zeta_1)R^{+(12)}(\zeta_1/\zeta_2, p^{*1/2}). \tag{1.2.3}$$

Here $R^+(\zeta, p^{1/2})$ denotes Baxter's elliptic R-matrix in the multiplicative notation, whose elliptic nome is $p^{1/2}$. We also set $p^{*1/2} = p^{1/2}q^{-c}$ with c being the canonical central element in $\widehat{\mathfrak{sl}}_2$.[4] This relation should be treated as a formal Laurent series in ζ. The coefficients are then formal power series in $p^{1/2}$ and well defined in the $p^{1/2}$-adic topology.

Initially the use of $R^+(\zeta, p^{1/2})$ and $R^+(\zeta, p^{*1/2})$ in (1.2.3) was due to a consistency to the formulation of the sine-Gordon model, which is obtained as a continuum limit of the XYZ spin chain model, by Lukyanov [111]. However due to this the usual matrix tensor product of $L^+(\zeta)$ does not work as a comultiplication for $\mathscr{A}_{q,p}(\widehat{\mathfrak{sl}}_2)$. Later the quasi-Hopf formulation of $\mathscr{A}_{q,p}(\widehat{\mathfrak{sl}}_N)$ was obtained in [77] and clarified that the appearance of two elliptic nomes $p^{1/2}$ and $p^{*1/2}$ is nothing but a dynamical shift. The same paper also provided a co-algebra structure to $\mathscr{A}_{q,p}(\widehat{\mathfrak{sl}}_N)$ and made it the vertex type elliptic quantum group. See Sect. 1.2.4.1.

The face type elliptic quantum groups were developed in quite a different way. The first attempt was done by Lukyanov and Pugai [113]. Following the idea of [54], they formulated a free field realization of the eight-vertex SOS model [7]. In particular they derived the two types of vertex operators by brute force. As mentioned in Sect. 1.1.3, critical behavior of the model is described by the Virasoro minimal model. Hence their work can be regarded as a formulation of a deformation of the Virasoro algebra. In fact, their screening operators appearing in their realization of the vertex operators coincide with those obtained by Shiraishi,

[4] $R^+(\zeta, p^{*1/2})$ is often denoted by $R^{+*}(\zeta)$.

Kubo, Awata, and Odake, who found a deformation of the Virasoro algebra as a bosonization of Macdonald difference operators [147]. Soon after, Jimbo, Konno, and Miwa investigated a degeneration limit of $\mathcal{A}_{q,p}(\widehat{\mathfrak{sl}}_2)$ to the massless regime of the XXZ model and observed that the two types of vertex operators and the screening operators coincide with the same degeneration limit of those obtained by Lukyanov and Pugai. This suggests that the degeneration limit of the Drinfeld currents of $\mathcal{A}_{q,p}(\widehat{\mathfrak{sl}}_2)$, if exist, should coincide with the same limit of the screening currents of the deformed Virasoro algebra. Motivated by this, Konno constructed a higher level extension of the screening currents of the deformed Virasoro algebra as a q-analogue of the Feigin–Fuchs construction and found that they form a closed elliptic algebra $U_{q,p}(\widehat{\mathfrak{sl}}_2)$ [94]. There the screening currents are treated as elliptic analogues of the Drinfeld currents, i.e. generating functions of the Drinfeld generators [30],[5] satisfying the relations whose coefficients are given in terms of Jacobi's theta functions.

It should be stressed that $U_{q,p}(\widehat{\mathfrak{sl}}_2)$ is by construction a face type elliptic algebra not the vertex type $\mathcal{A}_{q,p}(\widehat{\mathfrak{sl}}_2)$. This point was further investigated in [76] by constructing the L-operator of $U_{q,p}(\widehat{\mathfrak{sl}}_2)$ and showing it satisfies the dynamical RLL-relation introduced by Felder [45], which will be discussed in the next subsection. In [76], $U_{q,p}(\widehat{\mathfrak{sl}}_2)$ was also extended to the types of any untwisted affine Lie algebra $\mathfrak{g}$ [76]. However at that time the co-algebra structure of $U_{q,p}(\mathfrak{g})$ had not yet been known. There it was "borrowed" from the quasi-Hopf formulation of the face type elliptic quantum group $\mathcal{B}_{q,\lambda}(\mathfrak{g})$, which is a subject in Sect. 1.2.4.2, under the assumption that $U_{q,p}(\mathfrak{g})$ provides the Drinfeld realization of the same elliptic algebra as $\mathcal{B}_{q,\lambda}(\mathfrak{g})$ tensored by the dual elements e^Q's to the dynamical parameter P in λ. See Remarks 2.3 and 2.4. Later the co-algebra structure of $U_{q,p}(\mathfrak{g})$ was formulated in [38, 96, 97, 101] as an H-Hopf algebroid, which was introduced by Etingof and Varchenko [36] and developed by Koelink and Rosengren [86]. A detailed structure will be discussed in Chap. 3.

1.2.3 The Dynamical Yang–Baxter Equation and Felder's Elliptic Quantum Group

The face type elliptic quantum group had another development. It was started in 1984, when Gervais and Neveu found a strange equation, which is now called the dynamical Yang–Baxter equation (DYBE), in a study of the Liouville field theory [62]. The DYBE is a similar equation to the YBE but for slightly different R-matrices called the dynamical R-matrices depending on certain extra parameters, i.e. the dynamical parameters. The dynamical parameters get shifts in certain rule by weights of the vectors on which the dynamical R-matrices act. Ten years later

[5] Analogues of the loop generators of affine Lie algebras, i.e. generators in the formulation as central extensions of loop algebras, are called the Drinfeld generators.

Felder found that DYBE is equivalent to the face type YBE known as the star-triangle equation for the Boltzmann weights $W\left(\begin{matrix} a & b \\ d & c \end{matrix}\middle| u\right)$ of the face type solvable lattice models (Fig. 2) [45, 46].

$$
\sum_g W\left(\begin{matrix} a & b \\ f & g \end{matrix}\middle| u_1\right) W\left(\begin{matrix} b & c \\ g & d \end{matrix}\middle| u_2\right) W\left(\begin{matrix} f & g \\ e & d \end{matrix}\middle| u_1 - u_2\right)
$$

$$
= \sum_g W\left(\begin{matrix} a & b \\ g & c \end{matrix}\middle| u_1 - u_2\right) W\left(\begin{matrix} a & g \\ f & e \end{matrix}\middle| u_2\right) W\left(\begin{matrix} g & c \\ e & d \end{matrix}\middle| u_1\right). \quad (1.2.4)
$$

For the model associated with an affine Lie algebra $\mathfrak{g}$, the height variables a, b, c, d are elements of the dominant integral weights and satisfy the admissible conditions that $a-b, a-d, b-c, d-c$ take values in the weights in the vector representation of $\mathfrak{g}$ [81]. In fact, for example for the case $\mathfrak{g} = \widehat{\mathfrak{sl}}(N, \mathbb{C})$, let us set $a_i = (a+\rho, \bar{\epsilon}_i)$ $(i = 1, \cdots, N)$, $\widetilde{a} = \{a_i\}$ and define $R(u, \widetilde{a})$ by

$$
R(u, \widetilde{a}) = \sum_{\substack{i,j,k,l \\ \bar{\epsilon}_i + \bar{\epsilon}_j = \bar{\epsilon}_k + \bar{\epsilon}_l}} W\left(\begin{matrix} a & a+\bar{\epsilon}_k \\ a+\bar{\epsilon}_j & a+\bar{\epsilon}_i+\bar{\epsilon}_l \end{matrix}\middle| u\right) E_{ik} \otimes E_{jl}, \quad (1.2.5)
$$

where $\bar{\epsilon}_j$ are the weights of the vector representation $(\pi_V, V = \oplus_{j=1}^N \mathbb{C} v_j)$ and E_{ij} denotes the $N \times N$ matrix unit given by $E_{ij} v_k = \delta_{j,k} v_i$, then the face type YBE (1.2.4) can be rewritten as

$$
R^{(12)}(u_1 - u_2, \widetilde{a} + h^{(3)}) R^{(13)}(u_1, \widetilde{a}) R^{(23)}(u_2, \widetilde{a} + h^{(1)})
$$

$$
= R^{(23)}(u_2, \widetilde{a}) R^{(13)}(u_1, \widetilde{a} + h^{(2)}) R^{(12)}(u_1 - u_2, \widetilde{a}), \quad (1.2.6)
$$

where $h^{(l)}$ denotes the set $\{h_i^{(l)}$ $(i = 1, \cdots, N)\}$ $(l = 1, 2, 3)$, and $h_i v_j = \langle \bar{\epsilon}_j, h_i \rangle v_j$. See Sect. 1.2.4.2 for notations. The last equation is nothing but the DYBE.

Based on the DYBE, Felder formulated the face type elliptic algebra $E_{\tau,\eta}(\mathfrak{gl}_N)$ as an associative algebra generated by the dynamical L-operator $L(u, \widetilde{a})$ satisfying the *dynamical RLL-relation* [45, 46].

$$
R^{(12)}(u_1 - u_2, \widetilde{a} + h^{(3)}) L^{(1)}(u_1, \widetilde{a}) L^{(2)}(u_2, \widetilde{a} + h^{(1)})
$$

$$
= L^{(2)}(u_2, \widetilde{a}) L^{(1)}(u_1, \widetilde{a} + h^{(2)}) R^{(12)}(u_1 - u_2, \widetilde{a}). \quad (1.2.7)
$$

He also found that the comultiplication is given by

$$
\Delta(L(u, \widetilde{a})) = L^{(12)}(u, \widetilde{a} + h^{(3)}) \otimes L^{(13)}(u, \widetilde{a}), \quad (1.2.8)
$$

where $\otimes$ denotes the matrix tensor product. This is an elliptic and dynamical analogue of the FRST formulation of quantum groups.

However in [45], there were no description of the mode expansion of $L(u, \widetilde{a})$ in u so that the generators were unclear at all. In addition, $E_{\tau,\eta}(\mathfrak{gl}_N)$ does not have a canonical central element so it is an elliptic analogue of the quantum loop algebra $U_q(L\mathfrak{gl}_N)$, not of the quantum affine algebra $U_q(\widehat{\mathfrak{gl}}_N)$, in the FRST formulation. Later these points were resolved by Konno in [101] formulating the central extension $E_{q,p}(\widehat{\mathfrak{gl}}_N)$ of Felder's elliptic quantum group in a parallel way to $\mathscr{A}_{q,p}(\widehat{\mathfrak{gl}}_N)$ in [52]. See Appendix C. There to introduce the $RLL = LLR^*$ type relation is essential. It should be stressed that this type of relation is consistent to the formulation of $U_{q,p}(\widehat{\mathfrak{gl}}_N)$ as well as to $\mathscr{B}_{q,\lambda}(\widehat{\mathfrak{gl}}_N)$. See Remark 2.3. See also [61] for a different formulation without central extension, and [33, 34] for formulations with different central extensions.

As for the comultiplication (1.2.8), the dynamical shift sometimes becomes cumbersome, in particular when one needs to take it multiply. Concerning this point, the Hopf-algebroid structure introduced by Etingof and Varchenko [36], which we will discuss in Chap. 3, is useful even in the case with the central extension [96, 97]. The Hopf-algebroid structure was used in [87, 97] to derive the very-well-poised balanced elliptic hypergeometric series ${}_{12}V_{11}$ as a generalized elliptic $6j$-symbol. It was also shown in [96] the Hopf-algebroid structure is consistent with the quasi-Hopf formulation $\mathscr{B}_{q,\lambda}(\mathfrak{g})$ in a derivation of the vertex operators. See Chap. 5.

We also mention that the isomorphism between $E_{q,p}(\widehat{\mathfrak{gl}}_N)$ in the FRST formulation and $U_{q,p}(\widehat{\mathfrak{gl}}_N)$ in the Drinfeld realization was established in [101]. In Appendix C, we briefly discuss the $N = 2$ case.

1.2.4 Quasi-Hopf Formulation of the Elliptic Quantum Groups: $\mathscr{A}_{q,p}(\widehat{\mathfrak{sl}}_N)$ and $\mathscr{B}_{q,\lambda}(\mathfrak{g})$

Elliptic quantum groups have yet another formulation in terms of the Chevalley generators.

In the middle of 1990s, it was a big problem to find a general quantum group structure associated with the DYBE (1.2.6). An important work on this subject was done by Balelon, Bernard, and Billey [10]. They found that the dynamical constant R-matrix is obtained by a quasi-Hopf twist from the standard constant R-matrix of $U_q(\mathfrak{sl}_2)$. They also showed that the DYBE is a consequence of the same twist of the standard comultiplication and hence of the quasi-triangular structure of $U_q(\mathfrak{sl}_2)$. Here a key idea is to consider the quasi-Hopf deformation by a special twist operator called the shifted cocycle depending on the dynamical parameters. See below. Hence the notion of a dynamical quantum group was established as a quasi-Hopf deformation of a quasi-triangular Hopf algebra by the shifted cocycle.

The notion of a quasi-Hopf deformation of quantum group was introduced by Drinfeld in 1990 [31, 32]. Let $(A, \Delta, \varepsilon, S, \mathscr{R})$ be a quasi-triangular Hopf algebra consisting of a unital associative $\mathbb{C}$-algebra A, homomorphisms $\Delta : A \to A \otimes A$, $\varepsilon : A \to \mathbb{C}$, an antiautomorphism S, and the universal R-matrix $\mathscr{R}$. For an invertible element $\mathscr{T} \in A \times A$, he considered a "minimal" structure which generalizes the quasi-triangular Hopf algebra and is preserved by the twist operations

$$\widetilde{\Delta}(a) = \mathscr{T}\Delta(a)\mathscr{T}^{-1} \qquad (\forall a \in A), \tag{1.2.9}$$

$$\widetilde{\mathscr{R}} = \mathscr{T}^{(21)}\mathscr{R}\mathscr{T}^{(12)-1}. \tag{1.2.10}$$

He then found an algebraic structure called the quasi-triangular quasi-Hopf algebra $(A, \Delta, \varepsilon, S, \Phi, \mathscr{R}, \alpha, \beta)$. Here the new elements $\Phi \in A \otimes A \otimes A$ should be invertible and satisfy

$$(\mathrm{id} \otimes \Delta)\Delta(a) = \Phi(\Delta \otimes \mathrm{id})\Delta(a)\Phi^{-1} \quad \forall a \in A,$$

$$(\mathrm{id} \otimes \mathrm{id} \otimes \Delta)\Phi \cdot (\Delta \otimes \mathrm{id} \otimes \mathrm{id})\Phi = (1 \otimes \Phi) \cdot (\mathrm{id} \otimes \Delta \otimes \mathrm{id})\Phi \cdot (\Phi \otimes 1),$$

$$(\varepsilon \otimes \mathrm{id}) \circ \Delta = \mathrm{id} = (\mathrm{id} \otimes \varepsilon) \circ \Delta,$$

$$(\mathrm{id} \otimes \varepsilon \otimes \mathrm{id})\Phi = 1,$$

and $\alpha, \beta \in A$ are defined by the properties

$$\sum_i S(b_i)\alpha c_i = \varepsilon(a)\alpha, \qquad \sum_i b_i \beta S(c_i) = \varepsilon(a)\beta,$$

for $a \in A$, $\Delta(a) = \sum_i b_i \otimes c_i$, and

$$\sum_i X_i \beta S(Y_i) \alpha Z_i = 1,$$

for $\Phi = \sum_i X_i \otimes Y_i \otimes Z_i$. In addition, the axiom of the quasi-triangular structure is generalized as follows. There exists an invertible element $\mathscr{R} \in A \otimes A$ satisfying

$$\Delta'(a) = \mathscr{R}\Delta(a)\mathscr{R}^{-1}, \tag{1.2.11}$$

$$(\Delta \otimes \mathrm{id})\mathscr{R} = \Phi^{(312)}\mathscr{R}^{(13)}\Phi^{(132)^{-1}}\mathscr{R}^{(23)}\Phi^{(123)}, \tag{1.2.12}$$

$$(\mathrm{id} \otimes \Delta)\mathscr{R} = \Phi^{(231)^{-1}}\mathscr{R}^{(13)}\Phi^{(213)}\mathscr{R}^{(12)}\Phi^{(123)^{-1}}. \tag{1.2.13}$$

Here $\Delta' = \sigma \circ \Delta$, $\sigma(a \otimes b) = b \otimes a$ is the opposite comultiplication. In (1.2.12)–(1.2.13), if $\Phi = \sum_i X_i \otimes Y_i \otimes Z_i$, then we write $\Phi^{(312)} = \sum_i Z_i \otimes X_i \otimes Y_i$, $\Phi^{(213)} = \sum_i Y_i \otimes X_i \otimes Z_i$, and so forth. From (1.2.11)–(1.2.13) one obtains the Yang–Baxter type equation

$$\mathscr{R}^{(12)}\Phi^{(312)}\mathscr{R}^{(13)}\Phi^{(132)^{-1}}\mathscr{R}^{(23)}\Phi^{(123)} = \Phi^{(321)}\mathscr{R}^{(23)}\Phi^{(231)^{-1}}\mathscr{R}^{(13)}\Phi^{(213)}\mathscr{R}^{(12)}. \tag{1.2.14}$$

Note that the usual quasi-triangular Hopf algebra is a special case with $\Phi = 1$.

In fact, there are various twist operations which preserve the quasi-triangular quasi-Hopf-algebra structure. Let $(A, \Delta, \varepsilon, S, \Phi, \mathscr{R}, \alpha, \beta)$ be a quasi-triangular quasi-Hopf algebra. Let $\mathscr{T} \in A \otimes A$ be an invertible element such that $(\mathrm{id} \otimes \varepsilon)\mathscr{T} = 1 = (\varepsilon \otimes \mathrm{id})\mathscr{T}$. We refer to the element $\mathscr{T}$ as *twistor*. Set (1.2.9)–(1.2.10) and

$$\tilde{\Phi} = \left(\mathscr{T}^{(23)}(\mathrm{id} \otimes \Delta)\mathscr{T}\right) \Phi \left(\mathscr{T}^{(12)}(\Delta \otimes \mathrm{id})\mathscr{T}\right)^{-1}. \qquad (1.2.15)$$

Then $(A, \tilde{\Delta}, \varepsilon, S, \tilde{\Phi}, \tilde{\mathscr{R}}, \tilde{\alpha}, \tilde{\beta})$ with

$$\tilde{\alpha} = \sum_i S(d_i)\alpha e_i, \qquad \tilde{\beta} = \sum_i f_i \beta S(g_i)$$

is also a quasi-triangular quasi-Hopf algebra. Here we set $\sum_i d_i \otimes e_i = \mathscr{T}^{-1}$ and $\sum_i f_i \otimes g_i = \mathscr{T}$.

A key idea of Babelon–Bernard–Billey [10] is to consider a twist of a quasi-triangular *Hopf* algebra $(A, \Delta, \varepsilon, \mathscr{R})$ by the *shifted cocycle*. Let H be an abelian subalgebra of A, with the product written additively.

Definition 1.2.1 A twistor $\mathscr{T}(\eta)$ depending on $\eta \in H$ is a shifted cocycle if it satisfies the relation

$$\mathscr{T}^{(12)}(\eta)\,(\Delta \otimes \mathrm{id})\,\mathscr{T}(\eta) = \mathscr{T}^{(23)}(\eta + h^{(1)})\,(\mathrm{id} \otimes \Delta)\,\mathscr{T}(\eta) \qquad (1.2.16)$$

for some $h \in H$.

Let $(A, \Delta_\eta, \varepsilon, \Phi(\eta), \mathscr{R}(\eta))$ be the quasi-triangular quasi-Hopf algebra obtained from $(A, \Delta, \varepsilon, \mathscr{R})$ by a twist operation by $\mathscr{T}(\eta)$. The shifted cocycle condition (1.2.16) simplifies the properties of $\Phi(\eta)$ and $\mathscr{R}(\eta)$ as follows.

Proposition 1.2.1 *We have*

$$\Phi(\eta) = \mathscr{T}^{(23)}(\eta)\mathscr{T}^{(23)}(\eta + h^{(1)})^{-1}, \qquad (1.2.17)$$

$$(\Delta_\eta \otimes id)\mathscr{R}(\eta) = \Phi^{(312)}(\eta)\mathscr{R}^{(13)}(\eta)\mathscr{R}^{(23)}(\eta + h^{(1)}), \qquad (1.2.18)$$

$$(id \otimes \Delta_\eta)\mathscr{R}(\eta) = \mathscr{R}^{(13)}(\eta + h^{(2)})\mathscr{R}^{(12)}(\eta)\Phi^{(123)}(\eta)^{-1}. \qquad (1.2.19)$$

As a corollary the dynamical Yang–Baxter relation holds

$$\mathscr{R}^{(12)}(\eta + h^{(3)})\mathscr{R}^{(13)}(\eta)\mathscr{R}^{(23)}(\eta + h^{(1)}) = \mathscr{R}^{(23)}(\eta)\mathscr{R}^{(13)}(\eta + h^{(2)})\mathscr{R}^{(12)}(\eta). \qquad (1.2.20)$$

We hence reached the notion of a dynamical quantum group.

Definition 1.2.2 Let $(U_q(\mathfrak{g}), \Delta, \varepsilon, \mathscr{R})$ be a quantum group, i.e. a quasi-triangular Hopf algebra, in the Drinfeld–Jimbo formulation associated with a Kac–Moody Lie algebra $\mathfrak{g}$ [29, 73]. See Appendix A. Let $\mathscr{T}(\eta) \in U_q(\mathfrak{g}) \otimes U_q(\mathfrak{g})$ be a shifted cocycle with an element η in the Cartan subalgebra $\mathfrak{h}$ of $\mathfrak{g}$. We call the quasi-triangular quasi-Hopf algebra $(U_q(\mathfrak{g}), \Delta_\eta, \varepsilon, \Phi(\eta), \mathscr{R}(\eta))$ obtained from $(U_q(\mathfrak{g}), \Delta, \varepsilon, \mathscr{R})$ by a twist operation through $\mathscr{T}(\eta)$ the dynamical quantum group.

In order to formulate elliptic quantum groups, one needed one more step opened by Frønsdal [58, 59]. He considered quasi-Hopf deformations of the quantum *affine* algebra $U_q(\mathfrak{g})$, which hence *can* depend on the elliptic nome p, by shifted cocycles. See next two subsections. By constructing twistors a few terms in series expansions in p, he found that only two types of elliptic quasi-Hopf deformations of $U_q(\mathfrak{g})$ are allowed. These two types are distinguished by the automorphisms of $U_q(\mathfrak{g})$, and are nothing but the vertex and the face types. See below. Hence he reached an important observation that the vertex and the face type elliptic algebras are two different quasi-Hopf deformations of the quantum affine algebra $U_q(\mathfrak{g})$. Soon after a full construction of the twistors both of vertex type $\mathscr{E}(r)$ and face type $\mathscr{F}(\lambda)$ was done by Jimbo, Konno, Odake, and Shiraishi [77].

1.2.4.1 The Vertex Type Elliptic Quantum Group

Let $U = U_q(\widehat{\mathfrak{sl}}_N)$ and consider the affine quantum group $(U, \Delta, \varepsilon, \mathscr{R})$. For $r^* \in \mathbb{C}^\times$, introduce an automorphism of U by

$$\widetilde{\varphi}_{r^*} = \tau \circ \mathrm{Ad}\left(q^{\frac{2(r^*+c)}{N}\rho}\right). \tag{1.2.21}$$

Here τ denotes the Dynkin automorphism of $\widehat{\mathfrak{sl}}_N$ satisfying $\tau^N = \mathrm{id}$,

$$\tau(e_i) = e_{i+1 \bmod N}, \quad \tau(f_i) = f_{i+1 \bmod N}, \quad \tau(t_i) = t_{i+1 \bmod N}, \tag{1.2.22}$$

for the Chevalley generators e_i, f_i, t_i of $U_q(\mathfrak{g})$ in the Drinfeld–Jimbo formulation. See Appendix A, and c denotes the central element and $\rho \in \mathfrak{h}$ is an element such that $\langle \alpha_i, \rho \rangle = \frac{1}{2}(\alpha_i, \alpha_i)$ for all simple roots α_i $(i = 0, 1, \cdots, N-1)$. The adjoint operation Ad is defined by

$$\mathrm{Ad}(x)y = xyx^{-1}.$$

Set also

$$\widetilde{T} = \frac{1}{N}\left(\rho \otimes c + c \otimes \rho - \frac{N^2-1}{12} c \otimes c\right).$$

Then the vertex type twistor $\mathscr{E}(r^*)$ is given by

$$\mathscr{E}(r^*) = \prod_{k \geq 1}^{\curvearrowleft} \left(\widetilde{\varphi}_{r^*}^k \otimes \mathrm{id}\right) \left(q^{\widetilde{T}} \mathscr{R}\right)^{-1}. \tag{1.2.23}$$

Here $\prod_{k\geq 1}^{\curvearrowleft} A_k = \cdots A_3A_2A_1$, and the infinite product is to be understood as $\lim_{n\to\infty} \prod_{nN\geq k\geq 1}^{\curvearrowleft}$. Noting the commutation relations $[\rho, e_i] = e_i, [\rho, f_i] = -f_i$ $(i = 0, 1, \cdots, N-1)$, one finds that $\mathscr{E}(r^*)$ is a well-defined formal series in $p^{*\frac{1}{N}}$, where we set $p^* = q^{2r^*}$.

Then one obtains

Theorem 1.2.2 *The twistor* (1.2.23) *satisfies* $(\varepsilon \otimes id)\,\mathscr{E}(r^*) = (id \otimes \varepsilon)\,\mathscr{E}(r^*) = 1$ *and the shifted cocycle condition*

$$\mathscr{E}^{(12)}(r^*)(\Delta \otimes id)\mathscr{E}(r^*) = \mathscr{E}^{(23)}(r^* + c^{(1)})(id \otimes \Delta)\mathscr{E}(r^*). \tag{1.2.24}$$

We thus reach the following definition.

Definition 1.2.3 We define the vertex type elliptic quantum group $\mathscr{A}_{q,p}(\widehat{\mathfrak{sl}}_N)$ $(p = p^*q^{2c})$ to be the quasi-Hopf algebra $(U_q(\mathfrak{g}), \Delta_{r^*}, \varepsilon, \Phi(r^*), \mathscr{R}(r^*))$ obtained by the twist

$$\Delta_{r^*}(a) = \mathscr{E}^{(12)}(r^*)\,\Delta(a)\,\mathscr{E}^{(12)}(r^*)^{-1}, \tag{1.2.25}$$

$$\mathscr{R}(r^*) = \mathscr{E}^{(21)}(r^*)\,\mathscr{R}\,\mathscr{E}^{(12)}(r^*)^{-1}, \tag{1.2.26}$$

$$\Phi(r^*) = \mathscr{E}^{(23)}(r^*)\mathscr{E}^{(23)}(r^* + c^{(1)})^{-1}, \tag{1.2.27}$$

together with $\alpha_{r^*} = \sum_i S(d_i)e_i$, $\beta_{r^*} = \sum_i f_i S(g_i)$ and the antiautomorphism S defined by (A.1.7). Here $\sum_i d_i \otimes e_i = \mathscr{E}(r^*)^{-1}$, $\sum_i f_i \otimes g_i = \mathscr{E}(r^*)$.

It is also instructive to see that the vertex type elliptic algebra $\mathscr{A}_{q,p}(\widehat{\mathfrak{sl}}_N)$ in Sect. 1.2.2 is the same as defined here and is indeed a dynamical algebra. In fact the RLL-relation (1.2.3) is nothing but a representation $\pi_{V_{\zeta_1}} \otimes \pi_{V_{\zeta_2}} \otimes \mathrm{id}$ of the universal DYBE

$$\mathscr{R}^{(12)}(r^* + c^{(3)})\mathscr{R}^{(13)}(r^*)\mathscr{R}^{(23)}(r^* + c^{(1)}) = \mathscr{R}^{(23)}(r^*)\mathscr{R}^{(13)}(r^* + c^{(2)})\mathscr{R}^{(12)}(r^*).$$

Note that $p^{1/2} = p^{*1/2}q^c$, $\pi_{V_\zeta}(q^c) = 1$ and

$$R^+(\zeta_1/\zeta_2, p^{1/2}) = (\pi_{V_{\zeta_1}} \otimes \pi_{V_{\zeta_2}})q^{\widetilde{T}}\mathscr{R}(r^* + c),$$

$$R^+(\zeta_1/\zeta_2, p^{*1/2}) = (\pi_{V_{\zeta_1}} \otimes \pi_{V_{\zeta_2}})q^{\widetilde{T}}\mathscr{R}(r^*),$$

$$L^+(\zeta) = (\pi_{V_\zeta} \otimes \mathrm{id})q^{\widetilde{T}}\mathscr{R}(r^*).$$

See [77] for details.

1.2.4.2 The Face Type Elliptic Quantum Groups

Let $\mathfrak{g}$ be a symmetrizable Kac–Moody algebra and set $U = U_q(\mathfrak{g})$. Consider the associated quantum group $(U, \Delta, \varepsilon, \mathscr{R})$. See Appendix A. Let ϕ be an automorphism of U given by

$$\phi = \mathrm{Ad}(q^{\frac{1}{2}\sum_l h_l h^l - \rho}), \tag{1.2.28}$$

where $\{h_l\}$, $\{h^l\}$ are a basis and its dual basis of the Cartan subalgebra $\mathfrak{h}$, respectively. This yields

$$\phi(e_i) = e_i t_i, \qquad \phi(f_i) = t_i^{-1} f_i, \qquad \phi(q^h) = q^h.$$

For $\lambda \in \mathfrak{h}$, introduce an automorphism

$$\varphi_\lambda = \mathrm{Ad}(q^{\sum_l h_l h^l + 2(\lambda-\rho)}) = \phi^2 \circ \mathrm{Ad}(q^{2\lambda}). \tag{1.2.29}$$

Then the face type twistor $\mathscr{F}(\lambda)$ is given by

$$\mathscr{F}(\lambda) = \prod_{k\geq 1}^{\curvearrowleft} \left(\varphi_\lambda^k \otimes \mathrm{id}\right)\left(q^T \mathscr{R}\right)^{-1}. \tag{1.2.30}$$

Note that the k-th factor in the product (1.2.30) is a formal power series in the $q^{2k(\lambda,\alpha_i)}$ $(i \in I)$ with leading term 1, and hence the infinite product makes sense.

Then one can show the following.

Theorem 1.2.3 *The twistor* (1.2.30) *satisfies* $(\varepsilon \otimes id)\,\mathscr{F}(\lambda) = (id \otimes \varepsilon)\,\mathscr{F}(\lambda) = 1$ *and the shifted cocycle condition*

$$\mathscr{F}^{(12)}(\lambda)(\Delta \otimes \mathrm{id})\mathscr{F}(\lambda) = \mathscr{F}^{(23)}(\lambda + h^{(1)})(\mathrm{id} \otimes \Delta)\mathscr{F}(\lambda). \tag{1.2.31}$$

Hence one reaches the following definition.

Definition 1.2.4 We define the face type dynamical quantum group $\mathscr{B}_{q,\lambda}(\mathfrak{g})$ to be the quasi-Hopf algebra $(U_q(\mathfrak{g}), \Delta_\lambda, \varepsilon, \Phi(\lambda), \mathscr{R}(\lambda))$ obtained by the twist

$$\Delta_\lambda(a) = \mathscr{F}^{(12)}(\lambda)\,\Delta(a)\,\mathscr{F}^{(12)}(\lambda)^{-1}, \tag{1.2.32}$$

$$\mathscr{R}(\lambda) = \mathscr{F}^{(21)}(\lambda)\,\mathscr{R}\,\mathscr{F}^{(12)}(\lambda)^{-1}, \tag{1.2.33}$$

$$\Phi(\lambda) = \mathscr{F}^{(23)}(\lambda)\mathscr{F}^{(23)}(\lambda + h^{(1)})^{-1}, \tag{1.2.34}$$

together with $\alpha_\lambda = \sum_i S(d_i)e_i$, $\beta_\lambda = \sum_i f_i S(g_i)$ and the antiautomorphism S defined by (A.1.7). Here $\sum_i d_i \otimes e_i = \mathscr{F}(\lambda)^{-1}$, $\sum_i f_i \otimes g_i = \mathscr{F}(\lambda)$.

Remark 1.1 Note that the symmetrizable Kac–Moody algebra contains finite dimensional simple Lie algebras and affine Lie algebras as its major examples. It is important to note that *only* in the case when $\mathfrak{g}$ is of affine type, $\mathscr{B}_{q,\lambda}(\mathfrak{g})$ becomes elliptic quantum group and the universal R-matrix $\mathscr{R}(\lambda)$ gives a dynamical elliptic R-matrix [77, 95]. This is because only for an affine Lie algebra $\mathfrak{g}$ one can parametrize the dynamical parameter $\lambda \in \mathfrak{h}$ as

$$\lambda = (r^* + g)d + \sum_{i=1}^{l}(a_i + 1)\bar{h}^i, \tag{1.2.35}$$

where $r^*, a_i \in \mathbb{C}^\times$, g being the dual Coxeter number, and we take a dual basis $\{c, d, \bar{h}^1, \cdots, \bar{h}^l\}$ of $\mathfrak{h}$ with the scaling element d. Then the action $\mathrm{Ad}(q^{2\lambda})$ in (1.2.29) provides the elliptic nome $p^* = q^{2r^*}$ into $\mathscr{F}(\lambda)$ so as into $\mathscr{R}(\lambda)$ (1.2.33). Here we omit the term given by the central element c, because it does not contribute to $\mathscr{F}(\lambda)$. In this parameterization one obtains the known elliptic dynamical R-matrices $R^{+*}(z_1/z_2, \Pi^*)$ having the elliptic nome p^* by taking the vector representation of the universal R-matrix $(\pi_{z_1} \otimes \pi_{z_2})\mathscr{R}'^{+}(\lambda)$ [77, 95]. Here we set $\mathscr{R}'^{+}(\lambda) = q^{c\otimes d + d\otimes c}\mathscr{R}(\lambda)$ and introduced the multiplicative notations $z_i = q^{2u_i}$ $(i = 1, 2)$ and $\Pi_i^* = q^{2a_i}$ for the spectral and the dynamical parameters, respectively.[6] In addition, in the DYBE for $\mathscr{R}(\lambda)$

$$\mathscr{R}^{(12)}(\lambda + h^{(3)})\mathscr{R}^{(13)}(\lambda)\mathscr{R}^{(23)}(\lambda + h^{(1)}) = \mathscr{R}^{(23)}(\lambda)\mathscr{R}^{(13)}(\lambda + h^{(2)})\mathscr{R}^{(12)}(\lambda), \tag{1.2.36}$$

the dynamical shift $h^{(3)}$, for example, is parameterized as $h^{(3)} = c^{(3)}d + \sum_{i=1}^{l}\bar{h}_i^{(3)}\bar{h}^i$ similarly to (1.2.35). Hence we obtain

$$\lambda + h^{(3)} = (r^* + c^{(3)} + g)d + \sum_{i=1}^{l}(a_i + \bar{h}_i^{(3)} + 1)\bar{h}^i.$$

Therefore the shifted elliptic dynamical R-matrix $(\pi_{z_1} \otimes \pi_{z_2})\mathscr{R}'^{+}(\lambda + h^{(3)})$ becomes an elliptic R-matrix $R^{+}(z_1/z_2, \Pi^* q^{2\bar{h}})$ having the shift both in the elliptic nome $p \equiv p^* q^{2c}$ and in the dynamical parameter $\Pi^* q^{2\bar{h}}$. Then by taking the representation $\pi_{z_1} \otimes \pi_{z_2} \otimes \mathrm{id}$ of (1.2.36), one obtains the following dynamical RLL-relation.

$$\begin{aligned} R^{+(12)}(z_1/z_2, \Pi^* q^{2\bar{h}^{(3)}})L^{+(1)}(z_1, \Pi^*)L^{+(2)}(z_2, \Pi^* q^{2\bar{h}^{(1)}}) \\ = L^{+(2)}(z_2, \Pi^*)L^{+(1)}(z_1, \Pi^* q^{2\bar{h}^{(2)}})R^{+*(12)}(z_1/z_2, \Pi^*). \end{aligned} \tag{1.2.37}$$

[6] The symbol Π^* in R^{+*} denotes the set $\{\Pi_i^*\}$ $(i = 1, \cdots, \mathrm{rk}\bar{\mathfrak{g}})$. For example, for $\mathfrak{g} = \widehat{\mathfrak{sl}}_N$, Π_i^* appears in the matrix element $R^{+*}(z, \Pi^*)_{ij}^{kl}$ in the combination Π_i^*/Π_j^*. See [95, 101].

Here we set

$$L^+(z, \Pi^*) = (\pi_z \otimes \mathrm{id})\mathscr{R}'^+(\lambda).$$

The dynamical RLL-relation (1.2.7) used in Felder's formulation is a specialization $c = 0$, hence $p^* = p$, of (1.2.37) in the additive notation. There h should be read as $\bar{h}$ here.

Definition 1.2.5 For an affine Lie algebra $\mathfrak{g}$, we call $\mathscr{B}_{q,\lambda}(\mathfrak{g})$ the face type elliptic quantum group.

Remark 1.2 In [8], a similar construction of the face type twistor for the case $\mathfrak{g}$ being a finite dimensional simple Lie algebra was given. Hence dynamical quantum groups obtained in [8] are not elliptic ones. There a difference equation for the twistor, which is often referred as ABRR equation in literatures, was also obtained. The same difference equations for the twistors both of vertex type and face type were obtained in [77].

1.2.5 Summary

In summary one can classify the elliptic quantum groups into the following three formulations distinguished by their generators and co-algebra structures. See Table 1.1 : $\mathscr{A}_{q,p}(\widehat{\mathfrak{sl}}_N)$ and $\mathscr{B}_{q,\lambda}(\mathfrak{g})$ [77] in terms of the Chevalley generators, $U_{q,p}(\mathfrak{g})$ [76, 94] in terms of the Drinfeld generators, and $\mathscr{A}_{q,p}(\widehat{\mathfrak{sl}}_N)$ [52] and $E_{q,p}(\mathfrak{gl}_N)$ [36, 45, 47, 87, 101] in terms of the L operators. Here only $\mathscr{A}_{q,p}(\widehat{\mathfrak{sl}}_N)$ is the vertex type. The others are the face type. The co-algebra structures are the quasi-Hopf-algebra structure [31] for $\mathscr{A}_{q,p}(\widehat{\mathfrak{sl}}_N)$, $\mathscr{B}_{q,\lambda}(\mathfrak{g})$ [77], and the Hopf-algebroid structure [36, 86] for $E_{q,p}(\mathfrak{gl}_N)$ [47, 71, 87, 101] and $U_{q,p}(\mathfrak{g})$ [38, 97, 101].

As like in the rational and the trigonometric quantum groups, each formulation has both advantages and disadvantages. The quasi-Hopf-algebra formulations $\mathscr{A}_{q,p}(\widehat{\mathfrak{sl}}_N)$ and $\mathscr{B}_{q,\lambda}(\mathfrak{g})$ [77] are suitable for studying formal algebraic structures inherited from the trigonometric ones by the quasi-Hopf twist such as the universal elliptic dynamical R matrices, the universal form of the dynamical RLL relations as well as the existence of the intertwining operators (vertex operators), etc., but it

Table 1.1 Three formulations of the elliptic quantum groups

	Co-algebra structure	Generators
$\mathscr{A}_{q,p}(\widehat{\mathfrak{sl}}_N)$ (vertex type) $\mathscr{B}_{q,\lambda}(\mathfrak{g})$ (face type)	Quasi-Hopf algebra	Chevalley
$E_{q,p}(\mathfrak{g})$ (face type)	Hopf Algebroid	L-operator
$U_{q,p}(\mathfrak{g})$ (face type)	Hopf Algebroid	Drinfeld

is in general hard to derive concrete representations due to the complexity in taking representations of the twistors.

The FRST formulation $E_{q,p}(\widehat{\mathfrak{gl}}_N)$ is suitable for studying finite dimensional representations by a fusion procedure or by taking a comultiplication. In fact, finite dimensional representations of $E_{q,p}(\widehat{\mathfrak{gl}}_N)$ have been studied well [47, 87, 88] (see also [74]) and applied to the elliptic Ruijsenaars models [48, 49], the elliptic hypergeometric series [87, 88, 139], the partition function of the solvable lattice model [131, 140], and the elliptic Gaudin model [142].

The Drinfeld realization $U_{q,p}(\mathfrak{g})$ is suitable for studying both finite- and infinite dimensional representations [76, 90, 91, 93, 96, 97, 103] due to the nature of the Drinfeld generators as analogue of the loop generators of $\mathfrak{g}$. Recent developments include a characterization of the finite dimensional representations in terms of the theta function analogue [97] of the Drinfeld polynomials [22, 23, 30] and a clarification of the quantum Z-algebra structures of the infinite dimensional representations [38]. An application to the algebraic analysis of the solvable lattice models also has made a great success [19, 76, 78, 93, 96]. See also rather older works [6, 109, 113, 121] whose results, in particular the vertex operators and the screening operators, are able to be reformulated by the representation theory of $U_{q,p}(\widehat{\mathfrak{sl}}_N)$ [90, 93]. In addition as already mentioned in Sect. 1.1.3 there are deep relationships between $U_{q,p}(\mathfrak{g})$ and the deformed algebras $W_{p,p^*}(\bar{\mathfrak{g}})$: the generating functions of the Drinfeld generators (the elliptic currents) $e_j(z)$ and $f_j(z)$ of $U_{q,p}(\mathfrak{g})$ at level 1 are identified with the screening currents of the deformed W-algebras $W_{p,p^*}(\bar{\mathfrak{g}})$ of the coset type [38, 76, 92, 94]. It is also remarkable that the Drinfeld realization is suitable to formulate elliptic analogues of the quantum toroidal algebras [102, 104].

1.3 Notations

1.3.1 q-Integers, Infinite Products, and Theta Functions

For $p, q, t \in \mathbb{C}^\times$, $|q| < 1$, $|p| < 1$, $|t| < 1$, define

$$[n]_q := \frac{q^n - q^{-n}}{q - q^{-1}},$$

$$(z; q)_\infty := \prod_{n=0}^{\infty}(1 - zq^n), \quad (z; q, p)_\infty = \prod_{n,m=0}^{\infty}(1 - zq^n p^m).$$

In general we set

$$(z; q_1, q_2, \cdots, q_m)_\infty := \prod_{n_1, n_2, \cdots, n_m = 0}^{\infty}(1 - zq_1^{n_1} q_2^{n_2} \cdots q_m^{n_m}).$$

The multiple gamma functions are defined by

$$\Gamma(x;q,p)=\frac{(qp/x;q,p)_\infty}{(x;q,p)_\infty},$$
$$\Gamma(x;q,p,t)=(qpt/x;q,p,t)_\infty(x;q,p,t)_\infty.$$

In particular $\Gamma(x;q,p)$ is the elliptic Gamma function introduced by Ruijsenaars [143].

1.3.2 Theta Functions

Set

$$\theta(z,p):=-z^{-1/2}(z;p)_\infty(p/z;p)_\infty(p;p)_\infty.$$

The basic properties are

$$\theta(pz,p)=-p^{-1/2}z^{-1}\theta(z,p),\quad \theta(e^{2\pi i}z,p)=-\theta(z,p), \tag{1.3.1}$$
$$\theta(1/z,p)=-\theta(z,p). \tag{1.3.2}$$

We also have the identity

$$\theta(xy,p)\theta(x/y,p)\theta(zw,p)\theta(z/w,p)+\theta(xw,p)\theta(x/w,p)\theta(yz,p)\theta(y/z,p)$$
$$=\theta(xz,p)\theta(x/z,p)\theta(yw,p)\theta(y/w,p).$$

The elliptic gamma function satisfies

$$\Gamma(qz;q,p)=-z^{1/2}\frac{\theta(z,q)}{(q;q)_\infty}\Gamma(z;q,p),$$
$$\Gamma(pz;q,p)=-z^{1/2}\frac{\theta(z,p)}{(p;p)_\infty}\Gamma(z;q,p).$$

It may be instructive to make a connection to Jacobi odd theta function

$$\vartheta_1(v|\tau)=i\sum_{n\in\mathbb{Z}}(-1)^n e^{\pi i\tau(n-1/2)^2}e^{2\pi iv(n-1/2)}. \tag{1.3.3}$$

For $r\in\mathbb{C}^\times$, setting $p=q^{2r}=e^{-2\pi i/\tau}$, $|p|<1$, one finds

$$q^{u^2/r}\theta(q^{2u},p)=q^{-r/4}e^{-\pi i/4}\,\tau^{1/2}\vartheta_1(u/r|\tau).$$

Chapter 2
Elliptic Quantum Group $U_{q,p}(\widehat{\mathfrak{sl}}_2)$

In this chapter, the elliptic dynamical quantum group $U_{q,p}(\widehat{\mathfrak{sl}}_2)$ is defined by generators and relations. The generators are the Drinfeld type, i.e. an analogue of the loop generators of the affine Lie algebra $\widehat{\mathfrak{sl}}_2$. We call their generating functions the elliptic currents. The dynamical nature of $U_{q,p}(\widehat{\mathfrak{sl}}_2)$ is realized by introducing the dynamical parameter P and considering a copy $H = \mathbb{C}P$ of the Cartan subalgebra $\bar{\mathfrak{h}} = \mathbb{C}h$. We take the field $\mathcal{M}_{H^*}$ of meromorphic functions on H^* as the basic coefficient field and make it not commutative to the other generators of $U_{q,p}(\widehat{\mathfrak{sl}}_2)$. We also introduce the half currents and construct the L^+-operator $L^+(z)$ in the Gauss decomposed form by taking the half currents as its Gauss coordinates. It is then shown that the $L^+(z)$ satisfies the dynamical RLL-relation. In addition, following the quasi-Hopf formulation $\mathcal{B}_{q,\lambda}(\widehat{\mathfrak{sl}}_2)$, we introduce the L^--operator and show that the difference between the $+$ and the $-$ half currents gives the elliptic currents of $U_{q,p}(\widehat{\mathfrak{sl}}_2)$. Furthermore a connection to Felder's formulation is shown by introducing the dynamical L-operators.

2.1 The Affine Lie Algebra $\widehat{\mathfrak{sl}}_2$

Let us start by recalling the affine Lie algebra $\widehat{\mathfrak{sl}}_2$ [83]. Let $\mathfrak{sl}_2 = \mathfrak{sl}(2,\mathbb{C}) = < e, f, h >$. The affine Lie algebra $\widehat{\mathfrak{sl}}_2'$ is the extension of the loop algebra $\mathfrak{sl}_2 \otimes \mathbb{C}[t, t^{-1}]$ by a one-dimensional center $\mathbb{C}c$. By using the invariant bilinear form $\langle , \rangle$ on $\mathfrak{sl}_2$, the Lie bracket of $\widehat{\mathfrak{sl}}_2'$ is given by

$$[x \otimes t^m, y \otimes t^n] = [x, y] \otimes t^{m+n} + mc\langle x, y\rangle \delta_{m+n,0} \qquad \forall x, y \in \mathfrak{sl}_2, \forall m, n \in \mathbb{Z},$$
$$[c, \widehat{\mathfrak{sl}}_2] = 0.$$

H. Konno, *Elliptic Quantum Groups*, SpringerBriefs in Mathematical Physics 37,
https://doi.org/10.1007/978-981-15-7387-3_2

It is sometimes convenient to extend $\widehat{\mathfrak{sl}}'_2$ further by d satisfying the commutation relations

$$[d, x \otimes t^m] = m(x \otimes t^m),$$
$$[c, d] = 0.$$

We denote such extension by $\widehat{\mathfrak{sl}}_2 = \widehat{\mathfrak{sl}}'_2 \oplus \mathbb{C}d$. Consider the extension of the invariant bilinear form on $\widehat{\mathfrak{sl}}_2$ by

$$\langle x \otimes t^m, y \otimes t^n \rangle = \delta_{m+n,0} \langle x, y \rangle,$$
$$\langle x \otimes t^m, c \rangle = 0 = \langle x \otimes t^m, d \rangle,$$
$$\langle c, d \rangle = 1,$$
$$\langle c, c \rangle = 0 = \langle d, d \rangle.$$

Then $\langle , \rangle$ is non-degenerate on $\widehat{\mathfrak{sl}}_2$.

Let $\bar{\mathfrak{h}} = \mathbb{C}h$ be the Cartan subalgebra of $\mathfrak{sl}_2$. Then $\mathfrak{h} = \bar{\mathfrak{h}} \oplus \mathbb{C}c \oplus \mathbb{C}d$ gives the Cartan subalgebra of $\widehat{\mathfrak{sl}}_2$. Let $\mathfrak{h}^*$ denote the dual space of $\mathfrak{h}$. The non-degeneracy of $\langle , \rangle$ yields the identification $\mathfrak{h} \cong \mathfrak{h}^*$. We denote by α, δ, Λ_0 the dual elements in $\mathfrak{h}^*$ to h, c, d, respectively. Hence we have

$$\langle \alpha, h \rangle = 2, \quad \langle \alpha, c \rangle = 0 = \langle \alpha, d \rangle,$$
$$\langle \delta, h \rangle = 0 = \langle \Lambda_0, h \rangle,$$
$$\langle \delta, c \rangle = 0 = \langle \Lambda_0, d \rangle,$$
$$\langle \Lambda_0, c \rangle = 1 = \langle \delta, d \rangle.$$

Set $\bar{\Lambda}_1 = \alpha/2$ and $\Lambda_1 = \Lambda_0 + \bar{\Lambda}_1$. We call Λ_0, Λ_1 the fundamental weights of $\widehat{\mathfrak{sl}}_2$ and $\bar{\Lambda}_1$ the classical part of Λ_0. We denote by $\mathcal{Q} = \mathbb{Z}\alpha$ the root lattice and by $\mathcal{P} = \mathbb{Z}\bar{\Lambda}_1$ the weight lattice.

We regard $\bar{\mathfrak{h}} \oplus \bar{\mathfrak{h}}^*$ as a Heisenberg algebra by

$$[h, \alpha] = 2, \qquad [h, h] = 0 = [\alpha, \alpha] \tag{2.1.1}$$

and their linear extension on $\bar{\mathfrak{h}} \oplus \bar{\mathfrak{h}}^*$.

2.2 Dynamical Parameter

Let us introduce the dynamical parameter P and its dual element Q with the pairing $\langle Q, P \rangle = 2$, $\langle Q, \mathfrak{h} \oplus \mathfrak{h}^* \rangle = 0 = \langle \mathfrak{h} \oplus \mathfrak{h}^*, P \rangle$. We set $H = \mathbb{C}P$ and $H^* = \mathbb{C}Q$. Set also $P_\alpha = lP,\ Q_\alpha = lQ$ for $\alpha = l\mathfrak{a}, l \in \mathbb{C}$. In particular, $P_{\bar{\epsilon}_1} = -P_{\bar{\epsilon}_2} = P/2$ and $Q_{\bar{\epsilon}_1} = -Q_{\bar{\epsilon}_2} = Q/2$.

Let $\mathcal{Q}_Q = \mathbb{Z}Q$ denote the dynamical counterpart of the root lattice $\mathcal{Q}$. We denote by $\mathbb{F} = \mathcal{M}_{H^*}$ the field of meromorphic functions on H^*. We regard a meromorphic function $g(P)$ of P as an element in $\mathbb{F}$ by $g(P)(\lambda) = g(\langle\lambda, P\rangle)$ for $\lambda \in H^*$.

As above we regard $(\bar{\mathfrak{h}} \oplus H) \oplus (\bar{\mathfrak{h}}^* \oplus H^*)$ as a Heisenberg algebra adding the following relations to (2.1.1).

$$[P, Q] = 2, \quad [P, P] = 0 = [Q, Q],$$
$$[P, \bar{\mathfrak{h}} \oplus \bar{\mathfrak{h}}^*] = 0 = [Q, \bar{\mathfrak{h}} \oplus \bar{\mathfrak{h}}^*]. \tag{2.2.1}$$

We also assume P and Q commute with c, d, δ, Λ_0.

Introduce the group rings $\mathbb{C}[\mathcal{Q}]$ and $\mathbb{C}[\mathcal{Q}_Q]$ of $\mathcal{Q}$ and $\mathcal{Q}_Q$, respectively. For $\beta \in \mathcal{Q}$, the elements of $\mathbb{C}[\mathcal{Q}]$ and $\mathbb{C}[\mathcal{Q}_Q]$ are denoted by e^β and e^{Q_β}, respectively. Note that for $\alpha, \beta \in \mathcal{Q}$, we have $e^{Q_\alpha}e^{Q_\beta} = e^{Q_\alpha + Q_\beta}$, $(e^{Q_\alpha})^{-1} = e^{-Q_\alpha}$, $e^0 = 1$, etc.

2.3 Definition of $U_{q,p}(\widehat{\mathfrak{sl}}_2)$

Let $\hbar \in \mathbb{C}\backslash\{2\pi i\mathbb{Q}\}$ and set $q = e^\hbar$. We assume $|q| < 1$. Let p be indeterminate.

Definition 2.3.1 The elliptic algebra $U_{q,p}(\widehat{\mathfrak{sl}}_2)$ is a topological algebra over $\mathbb{F}[[p]]$ generated by $e_m, f_m, \alpha_n, q^{h/2}, K$, $(m \in \mathbb{Z}, n \in \mathbb{Z}_{\neq 0})$, $\widehat{d}$ and the central elements $q^{c/2}$. Introduce the generating functions, which we call the elliptic currents,

$$e(z) = \sum_{m\in\mathbb{Z}} e_m z^{-m}, \quad f(z) = \sum_{m\in\mathbb{Z}} f_m z^{-m},$$

$$\psi^+(q^{-c/2}z) = q^{-h}K^2 \exp\left(-(q-q^{-1})\sum_{n>0}\frac{\alpha_{-n}}{1-p^n}z^n\right)\exp\left((q-q^{-1})\sum_{n>0}\frac{p^n\alpha_n}{1-p^n}z^{-n}\right),$$

$$\psi^-(q^{c/2}z) = q^h K^2 \exp\left(-(q-q^{-1})\sum_{n>0}\frac{p^n\alpha_{-n}}{1-p^n}z^n\right)\exp\left((q-q^{-1})\sum_{n>0}\frac{\alpha_n}{1-p^n}z^{-n}\right).$$

The defining relations are given as follows. For $g(P) \in \mathbb{F}$,

$$g(P)e(z) = e(z)g(P-2),\ g(P)f(z) = f(z)g(P),\ [g(P), \alpha_n] = 0, \tag{2.3.1}$$

$$g(P)K = Kg(P-1),\ [g(P), q^{h/2}] = 0, \tag{2.3.2}$$

$$q^{h/2}e(z)q^{-h/2} = qe(z),\ q^{h/2}f(z)q^{-h/2} = q^{-1}f(z),\ [q^{h/2}, \alpha_n] = 0, \tag{2.3.3}$$

$$[K, e(z)] = [K, f(z)] = [K, \alpha_n] = [K, q^{h/2}] = [K, \widehat{d}] = 0, \tag{2.3.4}$$

$$[\widehat{d}, g(P)] = 0,\ [\widehat{d}, \alpha_n] = n\alpha_n,\ [\widehat{d}, e(z)] = -z\frac{\partial}{\partial z}e(z),\ [\widehat{d}, f(z)] = -z\frac{\partial}{\partial z}f(z), \tag{2.3.5}$$

$$[\alpha_m, \alpha_n] = \delta_{m+n,0}\frac{[2m]_q[cm]_q}{m}\frac{1-p^m}{1-p^{*m}}q^{-cm}, \tag{2.3.6}$$

$$[\alpha_n, e(z)] = \frac{[2n]_q}{n}\frac{1-p^n}{1-p^{*n}}q^{-cn}z^n e(z), \tag{2.3.7}$$

$$[\alpha_n, f(z)] = -\frac{[2n]_q}{n}z^n f(z), \tag{2.3.8}$$

$$z_1\frac{(q^2z_2/z_1; p^*)_\infty}{(p^*q^{-2}z_2/z_1; p^*)_\infty}e(z_1)e(z_2) = -z_2\frac{(q^2z_1/z_2; p^*)_\infty}{(p^*q^{-2}z_1/z_2; p^*)_\infty}e(z_2)e(z_1), \tag{2.3.9}$$

$$z_1\frac{(q^{-2}z_2/z_1; p)_\infty}{(pq^2z_2/z_1; p)_\infty}f_i(z_1)f(z_2) = -z_2\frac{(q^{-2}z_1/z_2; p)_\infty}{(pq^2z_1/z_2; p)_\infty}f(z_2)f_i(z_1), \tag{2.3.10}$$

$$[e(z_1), f(z_2)] = \frac{1}{q-q^{-1}}\left(\delta(q^{-c}z_1/z_2)\psi^-(q^{c/2}z_2) - \delta(q^c z_1/z_2)\psi^+(q^{-c/2}z_2)\right), \tag{2.3.11}$$

where we set $p^* = pq^{-2c}$ and $\delta(z) = \sum_{n\in\mathbb{Z}} z^n$. We also denote by $U'_{q,p}(\widehat{\mathfrak{sl}}_2)$ the subalgebra obtained by removing $\widehat{d}$.

We treat the relations (2.3.5), (2.3.7)–(2.3.11) as formal Laurent series in z, z_1, z_2. All the coefficients in z's are well defined in the p-adic topology.[1]

The relation (2.3.1) and (2.3.2) indicates that the elliptic currents $e(z), f(z)$ and the generators α_n, K carry the P-weight $-2, 0, 0, -1$, respectively. One thus obtains

$$U_{q,p}(\widehat{\mathfrak{sl}}_2)/pU_{q,p}(\widehat{\mathfrak{sl}}_2) \cong (U_q(\widehat{\mathfrak{sl}}_2) \otimes_{\mathbb{C}} \mathbb{F})\sharp e^{-Q/2}\mathbb{C}[\mathcal{Q}_Q]$$

by

$$e_m \mapsto x_m^+ e^{-Q}, \quad f_m \mapsto x_m^-, \quad \alpha_n \mapsto a_n, \quad K \mapsto e^{-Q/2}.$$

Here $x_m^\pm, a_n$ denote the Drinfeld generators of the quantum affine algebra $U_q(\widehat{\mathfrak{sl}}_2)$ (see, for example, [79]), and the smash product $\sharp$ is defined by

$$g(P)a \otimes e^{Q_\alpha} \cdot f(P)b \otimes e^{Q_\beta} = g(P)f(P - \langle Q_\alpha, P\rangle)ab \otimes e^{Q_\alpha + Q_\beta}$$

for $a, b \in U_q(\widehat{\mathfrak{sl}}_2)$, and $f(P), g(P) \in \mathbb{F}$, $e^{Q_\alpha}, e^{Q_\beta} \in \mathbb{C}[\mathcal{Q}_Q]$.

[1] This topology is the same concept as the h-adic topology used in [29].

We sometimes consider the quotient of $U_{q,p}(\widehat{\mathfrak{sl}}_2)$ by the relation $p = q^{2r}$ for $r \in \mathbb{C}^\times$. Moreover, we consider the level-k representation of $U_{q,p}(\widehat{\mathfrak{sl}}_2)$, on which the central element $q^{c/2}$ takes a value $q^{k/2}$, $k \in \mathbb{C}$. We denote the resultant algebra of level k by $U_{q,p}(\widehat{\mathfrak{sl}}_2)_k$, where $p^* = pq^{-2k} = q^{2(r-k)}$. Then the parameter r^* in Sect. 1.2.4.2 can be identified with $r^* = r - k$. We assume $|q|, |p|, |p^*| < 1$.

Remark 2.1 In $U_{q,p}(\widehat{\mathfrak{sl}}_2)_k$, from (2.3.9), (2.3.10), one obtains

$$e(z_1)e(z_2) = \frac{\theta^*(q^2 z_1/z_2)}{\theta^*(q^{-2} z_1/z_2)} e(z_2)e(z_1),$$

$$f(z_1)f(z_2) = \frac{\theta(q^{-2} z_1/z_2)}{\theta(q^2 z_1/z_2)} f(z_2)f(z_1),$$

where we set $\theta(z) = \theta(z, p)$ and $\theta^*(z) = \theta(z, p^*)$.

Furthermore setting $z = q^{2u}$ and introducing formal expressions $z^{-(P-1)/r^*}$ and $z^{(P+h-1)/r}$, one can consider

$$E(z) = e(z)z^{-(P-1)/r^*}, \qquad F(z) = f(z)z^{(P+h-1)/r}. \tag{2.3.12}$$

Then from (2.3.1) and (2.3.3) one obtains

$$E(z_1)E(z_2) = \frac{[u_1 - u_2 + 1]^*}{[u_1 - u_2 - 1]^*} E(z_2)E(z_1),$$

$$F(z_1)F(z_2) = \frac{[u_1 - u_2 - 1]}{[u_1 - u_2 + 1]} F(z_2)F(z_1).$$

Here

$$[u] = \vartheta_1(u/r|\tau), \qquad [u]^* = \vartheta_1\left(u/r^*|\tau^*\right)$$

denote the Jacobi odd theta functions (1.3.3) with $p = e^{-2\pi i/\tau}$, $p^* = e^{-2\pi i/\tau^*}$, $z_i = q^{2u_i}$ $(i = 1, 2)$. This convention is used in papers [76, 94, 97, 99, 100]. One may wonder about the single-valuedness of the quantities associated with $E(z)$ and $F(z)$, such as the $E(z)$, $F(z)$ counterparts of the half currents (2.5.3)–(2.5.4) as well as solutions to the elliptic q-KZ equation which will be discussed in Chap. 8. However, it is remarkable that the single-valuedness is always guaranteed by the quasi-periodicities of $[u]$, $[u]^*$ appearing in those quantities instead of $\theta(z)$, $\theta^*(z)$. See, for example, [76, 99]. See also Remark 8.2 for further implication of the extra factors introduced in (2.3.12).

2.4 The Elliptic Dynamical R-Matrix of Type $\widehat{\mathfrak{sl}}_2$

Let us introduce the multiplicative notation of the dynamical parameter by $\Pi^* = q^{2P}$ and its shifted counterpart $\Pi = \Pi^* q^{2h}$.

Let $V = \mathbb{C}v_1 \oplus \mathbb{C}v_2$ be the 2-dimensional representation of $\mathfrak{sl}_2$, i.e. $e = E_{12}$, $f = E_{21}$, $h = E_{11} - E_{22}$ on V. Here E_{ij} denotes the 2×2 matrix unit satisfying $E_{ij}v_k = \delta_{j,k}v_i$. Let $\bar{\epsilon}_1 = \alpha/2 = -\bar{\epsilon}_2 \in \bar{\mathfrak{h}}^*$. Then $hv_l = \langle \bar{\epsilon}_l, h \rangle v_l$.

The elliptic dynamical R-matrix $R^+(z, \Pi^*) \in \mathrm{End}_{\mathbb{C}}(V \otimes V)$ is given by

$$R^+(z, \Pi^*) = \rho^+(z)\bar{R}(z, \Pi^*), \tag{2.4.1}$$

$$\bar{R}(z, \Pi^*) = \begin{pmatrix} 1 & 0 & 0 & 0 \\ 0 & b(z, \Pi^*) & c(z, \Pi^*) & 0 \\ 0 & \bar{c}(z, \Pi^*) & \bar{b}(z) & 0 \\ 0 & 0 & 0 & 1 \end{pmatrix}, \tag{2.4.2}$$

where

$$\rho^+(z) = q^{-\frac{1}{2}} \frac{\Gamma(z; p, q^4)\Gamma(q^4 z; p, q^4)}{\Gamma(q^2 z; p, q^4)^2}, \tag{2.4.3}$$

$$b(z, \Pi^*) = \frac{\theta(q^2\Pi^*)\theta(q^{-2}\Pi^*)}{\theta(\Pi^*)^2} \frac{\theta(z)}{\theta(q^2 z)}, \qquad \bar{b}(z) = \frac{\theta(z)}{\theta(q^2 z)}, \tag{2.4.4}$$

$$c(z, \Pi^*) = \frac{\theta(z\Pi^*)\theta(q^2)}{\theta(\Pi^*)\theta(q^2 z)}, \qquad \bar{c}(z, \Pi^*) = \frac{\theta(z\Pi^{*-1})\theta(q^2)}{\theta(\Pi^{*-1})\theta(q^2 z)}.$$

The matrix elements of $R^+(z, \Pi^*)$ are defined by

$$R^+(z, \Pi^*) v_i \otimes v_j = \sum_{k,l} R^+(z, \Pi^*)_{kl}^{ij} v_k \otimes v_l. \tag{2.4.5}$$

The $R^+(z, \Pi^*)$ satisfies the DYBE

$$\begin{aligned} &R^{+(12)}(z_1/z_2, \Pi^* q^{2h^{(3)}}) R^{+(13)}(z_1/z_3, \Pi^*) R^{+(23)}(z_2/z_3, \Pi^* q^{2h^{(1)}}) \\ &= R^{+(23)}(z_2/z_3, \Pi^*) R^{+(13)}(z_1/z_3, \Pi^* q^{2h^{(2)}}) R^{+(12)}(z_1/z_2, \Pi^*) \end{aligned} \tag{2.4.6}$$

on $V \otimes V \otimes V$ and the unitarity

$$R^+(z, \Pi^*) R^{+(21)}(1/z, \Pi^*) = \mathrm{id}_{V \otimes V}. \tag{2.4.7}$$

Here $R^{+(ij)}(z, \Pi^*)$ $(i, j = 1, 2, 3)$ denotes the R-matrix acting on the tensor of the i-th V and the j-th V by (2.4.5). In particular, $R^{+(21)}(z, \Pi^*) = \mathsf{P}R^+(z, \Pi^*)\mathsf{P}$ with $\mathsf{P}(v_k \otimes v_l) = v_l \otimes v_k$. The element $q^{2h^{(i)}}$ acts on the i-th V by $q^{2h^{(i)}}v_l = q^{2\langle\bar{\epsilon}_l, h\rangle}v_l$.

Exercise 2.1 Show that the R-matrix (2.4.1) satisfies the zero-weight condition

$$[R^+(z, \Pi^*), h^{(1)} + h^{(2)}] = 0.$$

Exercise 2.2 Show the following formulas. Some of them are parts of (2.4.7).

$$(1)\ \rho^+(z)\rho^+(1/z) = 1 \tag{2.4.8}$$

$$(2)\ \rho^+(q^4z) = \frac{\theta(z)\theta(q^4z)}{\theta(q^2z)^2}\rho^+(z) \tag{2.4.9}$$

$$(3)\ b(z, \Pi^*)\bar{b}(1/z) + c(z, \Pi^*)c(1/z, \Pi^*) = 1 \tag{2.4.10}$$

$$(4)\ b(z, \Pi^*)\bar{b}(z) - c(z, \Pi^*)\bar{c}(z, \Pi^*) = \frac{\bar{b}(z)}{\bar{b}(1/z)} \tag{2.4.11}$$

Remark 2.2 The elliptic dynamical R-matrix (2.4.1) is nothing but the Boltzmann weight $W\left(\begin{matrix} a & b \\ c & d \end{matrix}\middle| z\right)$ for $a, b, c, d \in \mathfrak{h}$ of the 8-vertex SOS model [7, 81] through

$$R^+(z, q^{2P})_{ij}^{kl} = W\left(\begin{matrix} a & a + h_{\bar{\epsilon}_k} \\ a + h_{\bar{\epsilon}_j} & a + h_{\bar{\epsilon}_i} + h_{\bar{\epsilon}_j} \end{matrix}\middle| z\right)\delta_{\bar{\epsilon}_i+\bar{\epsilon}_j, \bar{\epsilon}_k+\bar{\epsilon}_l}.$$

Here we made the identification of the dynamical parameter $P = \langle\bar{\epsilon}_1 - \bar{\epsilon}_2, a\rangle + 1$ and set $h_{\bar{\epsilon}_1} = -h_{\bar{\epsilon}_2} = h/2$. The expression (2.4.1) can be obtained by taking the vector representation of the face type universal elliptic dynamical R-matrix (1.2.33) up to a gauge transformation [77, 95, 101].

2.5 The Half Currents and the L^+-Operator

Next let us define the half currents and the L-operator of $\mathcal{U}_k = U_{q,p}(\widehat{\mathfrak{sl}}_2)_k$.

Definition 2.5.1 (Half Currents) For $k \in \mathbb{C}$, we define the half currents $k_i^+(z)$ $(i = 1, 2)$, $e^+(z)$, $f^+(z) \in \mathcal{U}_k[[z, z^{-1}]]$ of $\mathcal{U}_k$ by

$$\begin{aligned} k_1^+(z) &= \exp\left\{-(q - q^{-1})\sum_{n>0}\frac{[n]_q}{[2n]_q}\frac{1}{1 - p^n}\alpha_{-n}(q^{-1}z)^n\right\} \\ &\quad \times \exp\left\{(q - q^{-1})\sum_{n>0}\frac{[n]_q}{[2n]_q}\frac{p^n}{1 - p^n}\alpha_n(q^{-1}z)^{-n}\right\}Kq^{-h/2}, \end{aligned} \tag{2.5.1}$$

$$
\begin{aligned}
k_2^+(z) &= k_1^+(q^2z)^{-1} \\
&= \mathcal{N} \exp\left\{(q-q^{-1})\sum_{n>0}\frac{[n]_q}{[2n]_q}\frac{1}{1-p^n}\alpha_{-n}(qz)^n\right\} \\
&\times \exp\left\{-(q-q^{-1})\sum_{n>0}\frac{[n]_q}{[2n]_q}\frac{p^n}{1-p^n}\alpha_n(qz)^{-n}\right\} K^{-1}q^{h/2}, \quad (2.5.2)
\end{aligned}
$$

$$
e^+(z) = a^* \oint_{C^*} \frac{dt}{2\pi i t} e(t) \frac{\theta^*(\Pi^{*-1}q^{2-k}z/t)\theta^*(q^2)}{\theta^*(q^{-k}z/t)\theta^*(\Pi^{*-1}q^2)}, \tag{2.5.3}
$$

$$
f^+(z) = a \oint_C \frac{dt}{2\pi i t} f(t) \frac{\theta(\Pi q^{-2}z/t)\theta(q^2)}{\theta(z/t)\theta(\Pi q^{-2})}. \tag{2.5.4}
$$

Here the contours C^* and C are defined by

$$
C^* : |q^{-k}z| < |t| < |p^{*-1}q^{-k}z|, \quad C : |z| < |t| < |p^{-1}z|.
$$

The contour integral picks up residues at the simple poles inside of these contours arising from $\theta^*(q^{-k}z/t)$ for (2.5.3) and from $\theta(z/t)$ for (2.5.4), respectively. The constants $\mathcal{N}, a, a^*$ are chosen as

$$
\mathcal{N} = \frac{(pq^2; p, q^4)_\infty^2}{(pq^4; p, q^4)_\infty (p; p, q^4)_\infty} \frac{(p^*q^4; p^*, q^4)_\infty (p^*; p^*, q^4)_\infty}{(p^*q^2; p^*, q^4)_\infty^2},
$$

$$
a^* = \frac{(p^*; p^*)_\infty}{(p^*q^2; p^*)_\infty}, \qquad a = \frac{(p; p)_\infty}{(pq^{-2}; p)_\infty}.
$$

It is easy to verify the following factorization formula.

Proposition 2.5.1

$$
\psi^+(q^{-k/2}z) = \varsigma\, k_1^+(z)k_2^+(z)^{-1}, \tag{2.5.5}
$$

where

$$
\varsigma = \frac{(p; p)_\infty (p^*q^2; p^*)_\infty}{(p^*; p^*)_\infty (pq^2; p)_\infty}.
$$

Note that the constants a, a^* and ς satisfy

$$
\frac{a^* a\, \varsigma\theta(q^2)}{(q-q^{-1})(p; p)_\infty^3} = 1. \tag{2.5.6}
$$

Proposition 2.5.2 ([76, 90]) *The half currents $e^+(z)$, $f^+(z)$ and $k_l^+(z)$ $(l = 1, 2)$ satisfy the following relations.*

$$k_l^+(z_1)k_l^+(z_2) = \rho(z)k_l^+(z_2)k_l^+(z_1), \tag{2.5.7}$$

$$k_1^+(z_2)k_2^+(z_1) = \rho(z)\frac{\bar{b}(z)}{\bar{b}^*(z)}k_2^+(z_1)k_1^+(z_2), \tag{2.5.8}$$

$$k_2^+(z_1)^{-1}e^+(z_2)k_2^+(z_1) = e^+(z_2)\frac{1}{\bar{b}^*(z)} - e^+(z_1)\frac{c^*(z, \Pi^*)}{\bar{b}^*(z)}, \tag{2.5.9}$$

$$k_2^+(z_1)f^+(z_2)k_2^+(z_1)^{-1} = \frac{1}{\bar{b}(z)}f^+(z_2) - \frac{\bar{c}(z, \Pi)}{\bar{b}(z)}f^+(z_1), \tag{2.5.10}$$

$$\begin{aligned}
&\frac{1}{\bar{b}^*(1/z)}e^+(z_1)e^+(z_2) - e^+(z_2)^2\frac{c^*(1/z, \Pi^* q^{-4})}{\bar{b}^*(1/z)} \\
&= \frac{1}{\bar{b}^*(z)}e^+(z_2)e^+(z_1) - e^+(z_1)^2\frac{c^*(z, \Pi^* q^{-4})}{\bar{b}^*(z)},
\end{aligned} \tag{2.5.11}$$

$$\begin{aligned}
&\frac{1}{\bar{b}(z)}f^+(z_1)f^+(z_2) - f^+(z_1)^2\frac{\bar{c}(z, \Pi q^{-2})}{\bar{b}(z)} \\
&= \frac{1}{\bar{b}(1/z)}f^+(z_2)f^+(z_1) - f^+(z_2)^2\frac{\bar{c}(1/z, \Pi q^{-2})}{\bar{b}(1/z)},
\end{aligned} \tag{2.5.12}$$

$$[e^+(z_1), f^+(z_2)] = k_1^+(z_2)k_2^+(z_2)^{-1}\frac{\bar{c}^*(z, \Pi^* q^{-2})}{\bar{b}^*(z)} - k_2^+(z_1)^{-1}k_1^+(z_1)\frac{\bar{c}(z, \Pi q^{-2})}{\bar{b}(z)}, \tag{2.5.13}$$

where $z = z_1/z_2$, and

$$\rho(z) = \frac{\rho^{+*}(z)}{\rho^+(z)}. \tag{2.5.14}$$

Proof

Eq. (2.5.7): From (2.3.6) and the formula $e^A e^B = e^{[A,B]}e^B e^A$ for A, B satisfying $[A, [A, B]] = 0 = [B, [A, B]]$, we have

$$\begin{aligned}
&e^{(q-q^{-1})\sum_{m>0}\frac{[m]_q}{[2m]_q}\frac{p^m}{1-p^m}\alpha_m(q^{-1}z_1)^{-m}}\, e^{-(q-q^{-1})\sum_{n>0}\frac{[n]_q}{[2n]_q}\frac{1}{1-p^n}\alpha_{-n}(q^{-1}z_2)^n} \\
&= f(z_1/z_2)e^{-(q-q^{-1})\sum_{n>0}\frac{[n]_q}{[2n]_q}\frac{1}{1-p^n}\alpha_{-n}(q^{-1}z_2)^n}\, e^{(q-q^{-1})\sum_{m>0}\frac{[m]_q}{[2m]_q}\frac{p^m}{1-p^m}\alpha_m(q^{-1}z_1)^{-m}},
\end{aligned}$$

where

$$f(z) = \exp\left\{-(q-q^{-1})^2 \sum_{m>0} \frac{1}{m} \frac{[m]_q^2[km]_q}{[2m]_q} \frac{p^m q^{-km}}{(1-p^m)(1-p^{*m})} z^{-m}\right\}$$
$$= \frac{(p^*q^4/z; p^*, q^4)_\infty (p^*/z; p^*, q^4)_\infty}{(p^*q^2/z; p^*, q^4)^2_\infty} \frac{(pq^2/z; p, q^4)^2_\infty}{(pq^4/z; p, q^4)_\infty (p/z; p, q^4)_\infty}$$

for $|p^*|, |p| < |z|$. In the second equality we used

$$q^{km} - q^{-km} = (1-p^{*m})q^{km} - (1-p^m)q^{-km}.$$

Hence we obtain

$$k_1^+(z_1)k_1^+(z_2) = f(z) : k_1^+(z_1)k_1^+(z_2) : .$$

Here : : denotes the normal ordering defined by

$$: \alpha_m \alpha_n := \begin{cases} \alpha_m \alpha_n \text{ if } m \le n \\ \alpha_n \alpha_m \text{ if } m > n \end{cases}.$$

Combining this with

$$k_1^+(z_2)k_1^+(z_1) = f(z_2/z_1) : k_1^+(z_1)k_1^+(z_2) : \quad \text{for } |z_1/z_2| < |p^{-1}|, |p^{*-1}|,$$

we obtain (2.5.7) for $l = 1$. The case $l = 2$ as well as (2.5.8) can be verified similarly.

Eq. (2.5.9): From (2.3.7) and the formula $e^A B e^{-A} = B + [A, B] + \frac{1}{2!}[A, [A, B]] + \cdots$, we have

$$e^{-(q-q^{-1})\sum_{n>0} \frac{[n]_q}{[2n]_q} \frac{1}{1-p^n} \alpha_{-n}(qz)^n} e(t) e^{(q-q^{-1})\sum_{n>0} \frac{[n]_q}{[2n]_q} \frac{1}{1-p^n} \alpha_{-n}(qz)^n}$$
$$= \frac{(q^{2-k}z/t; p^*)_\infty}{(q^{-k}z/t; p^*)_\infty} e(t),$$
$$e^{(q-q^{-1})\sum_{n>0} \frac{[n]_q}{[2n]_q} \frac{p^n}{1-p^n} \alpha_n(qz)^{-n}} e(t) e^{-(q-q^{-1})\sum_{n>0} \frac{[n]_q}{[2n]_q} \frac{p^n}{1-p^n} \alpha_n(qz)^{-n}}$$
$$= \frac{(p^*q^{-2+k}t/z; p^*)_\infty}{(p^*q^k t/z; p^*)_\infty} e(t)$$

for $|p^*q^{k-2}| < |z_1/z_2| < |q^k|$. Noting also $q^{-h/2}e(t)q^{h/2} = q^{-1}e(t)$, we obtain

$$k_2^+(z)^{-1}e(t)k_2^+(z) = \frac{\theta^*(q^{2-k}z/t)}{\theta^*(q^{-k}z/t)} e(t).$$

Hence the LHS of (2.5.9) is given by

$$k_2^+(z_1)^{-1}e^+(z_2)k_2^+(z_1) = a^* \oint_{C^*} \frac{dt}{2\pi i t} e(t) \frac{\theta^*(\Pi^{*-1}q^{-k}z_2/t)\theta^*(q^2)\theta^*(q^{2-k}z_1/t)}{\theta^*(q^{-k}z_2/t)\theta^*(\Pi^{*-1})\theta^*(q^{-k}z_1/t)},$$

where we used $\Pi^{*-1}k_2^+(z) = k_2^+(z)\Pi^{*-1}q^{-2}$. On the other hand, the RHS of (2.5.9), we have

$$a^* \oint_{C^*} \frac{dt}{2\pi i t} e(t) \left\{ \frac{\theta^*(\Pi^{*-1}q^{2+k}z_2/t)\theta^*(q^2)\theta^*(q^2 z_1/z_2)}{\theta^*(q^{-k}z_2/t)\theta^*(\Pi^{*-1}q^2)\theta^*(z_1/z_2)} \right.$$
$$\left. - \frac{\theta^*(\Pi^{*-1}q^{2-k}z_1/t)\theta^*(q^2)\theta^*(\Pi^* z_1/z_2)\theta^*(q^2)}{\theta^*(q^{-k}z_1/t)\theta^*(\Pi^{*-1}q^2)\theta^*(z_1/z_2)\theta^*(\Pi^*)}. \right\}$$

Then due to the identity

$$\begin{aligned}
&\theta^*(\Pi^{*-1}q^{2+k}z_2/t)\theta^*(q^2 z_1/z_2)\theta^*(q^{-k}z_1/t)\theta^*(\Pi^*)\\
&\quad -\theta^*(\Pi^{*-1}q^{2-k}z_1/t)\theta^*(q^2)\theta^*(q^{-k}z_2/t)\theta^*(\Pi^* z_1/z_2)\\
&= -\theta^*(\Pi^{*-1}q^{-k}z_2/t)\theta^*(q^{2-k}z_1/t)\theta^*(z_1/z_2)\theta^*(\Pi^{*-1}q^2),
\end{aligned}$$

one finds that the RHS coincides with the LHS.

Eq. (2.5.13): Using (2.3.11) and integrating the formal delta function by t_1, one finds that the LHS is

$$a^* a \oint_{C^*} \frac{dt_1}{2\pi i t_1} \oint_C \frac{dt_2}{2\pi i t_2} [e(t_1), f(t_2)] \frac{\theta^*(\Pi^{*-1}q^{-k+2}z_1/t_1)\theta^*(q^2)}{\theta^*(q^{-k}z_1/t_1)\theta^*(\Pi^{*-1}q^2)} \frac{\theta(\Pi q^{-2}z_2/t_2)\theta(q^2)}{\theta(z_2/t_2)\theta(\Pi q^{-2})}$$
$$= \frac{a^* a}{q-q^{-1}} \left\{ \oint_{C_-} \frac{dt_2}{2\pi i t_2} \psi^-(q^{k/2}t_2) \frac{\theta^*(\Pi^{*-1}q^{-2k+2}z_1/t_2)\theta(\Pi q^{-2}z_2/t_2)}{\theta^*(q^{-2k}z_1/t_2)\theta(z_2/t_2)} \right.$$
$$\left. - \oint_{C_+} \frac{dt_2}{2\pi i t_2} \psi^+(q^{-k/2}t_2) \frac{\theta^*(\Pi^{*-1}q^2 z_1/t_2)\theta(\Pi q^{-2}z_2/t_2)}{\theta^*(z_1/t_2)\theta(z_2/t_2)} \right\} \frac{\theta^*(q^2)\theta(q^2)}{\theta^*(\Pi^{*-1}q^2)\theta(\Pi q^{-2})}.$$

Here C_- circles the simple poles $q^{-2k}p^{*m}z_1, p^m z_2$, whereas C_+ circles $p^{*m}z_1, p^m z_2$ $(m \in \mathbb{Z}_{\geq 0})$. In the 1st term change the integration variable $t_2' = pt_2$ and use $\psi^-(p^{-1}q^{k/2}t_2') = q^{2h}\psi^+(q^{-k/2}t_2')$. Then one can combine the two terms into

$$-\frac{a^* a}{q-q^{-1}} \oint_{C'} \frac{dt_2}{2\pi i t_2} \psi^+(q^{-k/2}t_2) \frac{\theta^*(\Pi^{*-1}q^2 z_1/t_2)\theta(\Pi q^{-2}z_2/t_2)\theta^*(q^2)\theta(q^2)}{\theta^*(z_1/t_2)\theta(z_2/t_2)\theta^*(\Pi^{*-1}q^2)\theta(\Pi q^{-2})},$$

where C' circles z_1 and z_2. Picking the residues at these points and using (2.5.5), (2.5.8) and (2.5.6), one reaches the desired result.

The proof of (2.5.11) is left as an exercise for the reader. See [76]. □

Definition 2.5.2 (L-Operator) We define the L-operator $L^+(z) \in \mathrm{End}_{\mathbb{C}}(V) \otimes \mathscr{U}_k$ by

$$L^+(z) = \begin{pmatrix} 1 & f^+(z) \\ 0 & 1 \end{pmatrix} \begin{pmatrix} k_1^+(z) & 0 \\ 0 & k_2^+(z) \end{pmatrix} \begin{pmatrix} 1 & 0 \\ e^+(z) & 1 \end{pmatrix}. \tag{2.5.15}$$

Note that

$$L^+(z) = \begin{pmatrix} k_1^+(z) + f^+(z)k_2^+(z)e^+(z) & f^+(z)k_2^+(z) \\ k_2^+(z)e^+(z) & k_2^+(z) \end{pmatrix} \tag{2.5.16}$$

and

$$L^+(z)^{-1} = \begin{pmatrix} k_1^+(z)^{-1} & -k_1^+(z)^{-1}f^+(z) \\ -e^+(z)k_1^+(z)^{-1} & k_2^+(z)^{-1} + e^+(z)k_1^+(z)^{-1}f^+(z) \end{pmatrix}. \tag{2.5.17}$$

Then one can show the following statement.

Theorem 2.5.3 ([76]) *The $L^+(z)$ satisfies the dynamical RLL-relation*

$$R^{+(12)}(z_1/z_2, \Pi)L^{+(1)}(z_1)L^{+(2)}(z_2) = L^{+(2)}(z_2)L^{+(1)}(z_1)R^{+*(12)}(z_1/z_2, \Pi^*). \tag{2.5.18}$$

Here $R^{+*}(z, \Pi^*)$ denotes the same elliptic dynamical R-matrix as (2.4.1) *with replacement $\Pi \mapsto \Pi^*$ and $p \mapsto p^*$, hence $\theta(z) \mapsto \theta^*(z)$, etc., and*

$$L^{+(1)}(z) = L^+(z) \otimes \mathrm{id}, \qquad L^{+(2)}(z) = \mathrm{id} \otimes L^+(z).$$

Proof In the component form, (2.5.18) is given by

$$\sum_{i',j'} R^+(z_1/z_2, \Pi)_{ij}^{i'j'} L^+_{i'i''}(z_1)L^+_{j'j''}(z_2) = \sum_{i',j'} L^+_{jj'}(z_2)L^+_{ii'}(z_1)R^{+*}(z_1/z_2, \Pi^*)_{i'j'}^{i''j''}. \tag{2.5.19}$$

We call this the $(i, j), (i'', j'')$ component of (2.5.18). Note also (2.5.18) is equivalent to

$$L^{+(2)}(z_2)^{-1}R^{+(12)}(z_1/z_2, \Pi)L^{+(1)}(z_1) = L^{+(1)}(z_1)R^{+*(12)}(z_1/z_2, \Pi^*)L^{+(2)}(z_2)^{-1}, \tag{2.5.20}$$

$$L^{+(1)}(z_1)^{-1}L^{+(2)}(z_2)^{-1}R^{+(12)}(z_1/z_2, \Pi) = R^{+*(12)}(z_1/z_2, \Pi^*)L^{+(2)}(z_2)^{-1}L^{+(1)}(z_1)^{-1}. \tag{2.5.21}$$

Then

(i) the $(1,1),(1,1)$ component of (2.5.21) and the $(2,2),(2,2)$ component of (2.5.18) are equivalent to (2.5.7).
(ii) the $(2,1),(2,1)$ component of (2.5.20) is equivalent to (2.5.8).
(iii) the (2,2), (2,1) component of (2.5.18) with (2.5.7) is equivalent to (2.5.9).
(iv) the (2,1), (2,2) component of (2.5.18) with (2.5.7) is equivalent to (2.5.10).
(v) from the (1,1), (2,2) component of (2.5.18), we have

$$f^+(z_1)k_2^+(z_1)f^+(z_2)k_2^+(z_2) = \rho(z_1/z_2)f^+(z_2)k_2^+(z_2)f^+(z_1)k_2^+(z_1). \tag{2.5.22}$$

Applying (2.5.7) and (2.5.10), we obtain (2.5.12). Similarly combining the $(2,2),(1,1)$ component of (2.5.18) with (2.5.9), one gets (2.5.11).
(vi) from the (2,2), (1,2) component of (2.5.18), we have

$$\begin{aligned}
&\rho^+(z_1/z_2)k_2^+(z_1)e^+(z_1)k_2^+(z_2)\\
&= \rho^{+*}(z_1/z_2)\left(k_2^+(z_2)k_2^+(z_1)e^+(z_1)b^*(z_1/z_2,\Pi^*)\right.\\
&\quad \left.+\, k_2^+(z_2)e^+(z_2)k_2^+(z_1)\bar{c}^*(z_1/z_2,\Pi^*)\right),
\end{aligned} \tag{2.5.23}$$

which is equivalent to (2.5.9) due to (2.4.11). From the (2,1), (1,2) component of (2.5.18), we have

$$\begin{aligned}
&\rho^+(z_1/z_2)\Big\{\bar{b}(z_1/z_2)k_2^+(z_1)e^+(z_1)f^+(z_2)k_2^+(z_2)\\
&\quad +\bar{c}(z_1/z_2,\Pi)\left(k_1^+(z_1)+f^+(z_1)k_2^+(z_1)e^+(z_1)\right)k_2^+(z_2)\Big\}\\
&= \rho^{*+}(z_1/z_2)\Big\{\left(k_1^+(z_2)+f^+(z_2)k_2^+(z_2)e^+(z_2)\right)k_2^+(z_1)\bar{c}^*(z_1/z_2,\Pi^*)\\
&\quad +f^+(z_2)k_2^+(z_2)k_2^+(z_1)e^+(z_1)b^*(z_1/z_2,\Pi^*)\Big\}.
\end{aligned}$$

Applying (2.5.9), (2.5.23) to this and using $\Pi^* k_2^+(z)^{-1} = k_2^+(z)^{-1}\Pi^* q^{-2}$, $\Pi k_1^+(z) = k_1^+(z)\Pi q^{-2}$, one gets (2.5.13). The other components of (2.5.18) are equivalent to one of the relations in Proposition 2.5.2. □

Remark 2.3 The dynamical RLL-relation (2.5.18) turns out consistent to the one (1.2.37) derived in the quasi-Hopf formulation $\mathscr{B}_{q,\lambda}(\widehat{\mathfrak{sl}}_2)$ in [77]. See also Sect. 2.7.

Exercise 2.3 Show

$$q^{P+h}L_{ij}^+(z)q^{-(P+h)} = q^{-\langle Q_{\bar{\epsilon}_i}, P+h\rangle}L_{ij}^+(z), \tag{2.5.24}$$

$$q^{P}L_{ij}^+(z)q^{-P} = q^{-\langle Q_{\bar{\epsilon}_j}, P\rangle}L_{ij}^+(z). \tag{2.5.25}$$

2.6 L^--Operator

For later convenience we next introduce $R^-(z,\Pi)$ and $L^-(z)=\sum_{i,j=1,2}E_{ij}L^-_{ij}(z)$. We follow the results in the quasi-Hopf formulation $\mathscr{B}_{q,\lambda}(\widehat{\mathfrak{sl}}_2)$ [77] and define them by

$$R^-(z,\Pi)=\rho^-(z)\bar{R}(z,\Pi),\qquad \rho^-(z)=q\rho^+(pz),\tag{2.6.1}$$

$$L^-(z)=\left(\mathrm{Ad}(q^{2\theta_V(P)})\otimes\mathrm{id}\right)\left(q^{2T_V}L^+(zp^*q^c)\right),\tag{2.6.2}$$

where

$$\theta_V(P)=\frac{1}{2}\pi_V(h)\pi_V(h_{\bar{\epsilon}_1})+P\pi_V(h_{\bar{\epsilon}_1}),\tag{2.6.3}$$

$$T_V=\pi_V(h)\otimes h_{\bar{\epsilon}_1},\tag{2.6.4}$$

and $\pi_V(h)=E_{11}-E_{22}$, $\pi_V(h_{\bar{\epsilon}_1})=\frac{1}{2}(E_{11}-E_{22})$. Then one can verify the following statement.

Proposition 2.6.1 *The L operators $L^+(z)$ and $L^-(z)$ satisfy the following relations.*

$$R^{-(12)}(z_1/z_2,\Pi)L^{-(1)}(z_1)L^{-(2)}(z_2)=L^{-(2)}(z_2)L^{-(1)}(z_1)R^{-*(12)}(z_1/z_2,\Pi^*),\tag{2.6.5}$$

$$R^{\pm(12)}(q^{\pm k}z_1/z_2,\Pi)L^{\pm(1)}(z_1)L^{\mp(2)}(z_2)=L^{\mp(2)}(z_2)L^{\pm(1)}(z_1)R^{\pm*(12)}(q^{\mp k}z_1/z_2,\Pi^*).\tag{2.6.6}$$

Proof Replace z_i with $z_ip^*q^k$ $(i=1,2)$ in (2.5.18). Note that (2.5.24), (2.5.25), and (2.6.2) yields

$$L^+(p^*q^kz)=\sum_{i,j}q^{-2(P+h)_{\bar{\epsilon}_i}}q^{2P_{\bar{\epsilon}_j}}E_{ij}L^-_{ij}(z).$$

By a componentwise comparison we obtain

$$R^+(z_1/z_2,\Pi)L^-(z_1)L^-(z_2)=L^-(z_2)L^-(z_1)R^{+*}(z_1/z_2,\Pi^*).$$

Then noting

$$\rho(z)=\frac{\rho^{+*}(z)}{\rho^+(z)}=\frac{\rho^{-*}(z)}{\rho^-(z)},$$

we obtain (2.6.5).

Similarly let us replace z_1 by $z_1 p^* q^k$ in (2.5.18). Noting $p^* q^k = pq^{-k}$, the components of R^+ are changed as

$$b(zpq^{-k}, \Pi) = q^2 b(zq^{-k}, \Pi), \qquad \bar{b}(zpq^{-k}) = q^2 \bar{b}(zq^{-k}), \tag{2.6.7}$$
$$c(zpq^{-k}, \Pi) = q^2 \Pi^{-1} c(zq^{-k}, \Pi), \qquad \bar{c}(zpq^{-k}, \Pi) = q^2 \Pi \bar{c}(zq^{-k}, \Pi)$$

and similarly for R^{+*}. Then from (2.4.1), we obtain the second (lower sign) relation in (2.6.6). Note that a factor arising from the action of $\mathrm{Ad}(q^{-2\theta_V(P)}) \otimes \mathrm{id}$ on the L-operators cancels the extra factors in (2.6.7).

To obtain the first relation in (2.6.6), exchange z_1 and z_2 in the second relation of (2.6.6). Then we have

$$R^-(q^{-k} z_2/z_1, \Pi)^{-1} L^+(z_1) L^-(z_2) = L^-(z_2) L^+(z_1) R^{-*}(q^k z_2/z_1, \Pi^*)^{-1}.$$

Using

$$R^-(z, \Pi)^{-1} = \mathsf{P} R^+(z^{-1}, \Pi) \mathsf{P},$$

one obtains the desired result. □

In addition, let us define $e^-(z), f^-(z), k_i^-(z)$ $(i = 1, 2)$ as the Gauss coordinates of $L^-(z)$ in the same way as (2.5.15). Then from the definition of $L^-(z)$ (2.6.2), we have

$$e^-(z) = q^{2P_{\bar{\epsilon}_2}} e^+(zpq^{-k}) q^{-2P_{\bar{\epsilon}_1}}, \tag{2.6.8}$$
$$f^-(z) = q^{2(P+h)_{\bar{\epsilon}_1}} f^+(zpq^{-k}) q^{-2(P+h)_{\bar{\epsilon}_2}}, \tag{2.6.9}$$
$$k_i^-(z) = qq^{2(P+h)_{\bar{\epsilon}_i}} k_i^+(zpq^{-k}) q^{-2P_{\bar{\epsilon}_i}} \tag{2.6.10}$$

for $i = 1, 2$. Then from (2.5.3)–(2.5.4) one can show the following statement.

Proposition 2.6.2

$$e^+(q^k z) - e^-(z) = -\frac{a^* \theta^*(q^2)}{(p^*; p^*)^3_\infty} e(z),$$

$$f^+(z) - f^-(q^k z) = -\frac{a \theta(q^2)}{(p; p)^3_\infty} f(z).$$

2.7 The Dynamical L-Operators and the Dynamical Half Currents

In order to obtain a "fully" dynamical RLL-relations used in [45, 47] with a central extension, one may introduce the dynamical L-operators $L^{\pm}(z,\Pi^*)$ related to our $L^{\pm}(z)$ by [76, 90]

$$L^{\pm}(z,\Pi^*)=L^{\pm}(z)e^{\sum_{i=1,2}\pi_V(h_{\epsilon_i})\otimes Q_{\bar{\epsilon}_i}}\in \mathrm{End}_{\mathbb{C}}(V)\otimes \mathscr{U}_k, \tag{2.7.1}$$

where $\pi_V(h_{\epsilon_i})=E_{ii}$. In fact from (2.5.24) and (2.5.25) we have

$$[L^{\pm}_{ij}(z,\Pi^*),g(P)]=0, \tag{2.7.2}$$

$$q^hL^{\pm}_{ij}(z,\Pi^*)=L^{\pm}_{ij}(z,\Pi^*)q^hq^{-\langle\bar{\epsilon}_i-\bar{\epsilon}_j,h\rangle}, \tag{2.7.3}$$

$$[\widehat{d},L^{\pm}(z,\Pi^*)]=-z\frac{\partial}{\partial z}L^{\pm}(z,\Pi^*). \tag{2.7.4}$$

Note that (2.7.2) indicates that $L^{\pm}(z,P)$ is independent of $e^{-Q/2}\mathbb{C}[\mathscr{Q}_Q]$. Furthermore from (2.2.1), (2.5.18), (2.6.5), and (2.6.6), $L^{\pm}(z,\Pi^*)$ satisfy the following full dynamical RLL-relations

$$\begin{aligned}&R^{\pm(12)}(z_1/z_2,\Pi)L^{\pm(1)}(z_1,\Pi^*)L^{\pm(2)}(z_2,\Pi^*q^{2\pi_V(h)^{(1)}})\\&\quad=L^{\pm(2)}(z_2,\Pi^*)L^{\pm(1)}(z_1,\Pi^*q^{2\pi_V(h)^{(2)}})R^{\pm*(12)}(z_1/z_2,\Pi^*),\end{aligned} \tag{2.7.5}$$

$$\begin{aligned}&R^{\pm(12)}(q^{\pm c}z_1/z_2,\Pi)L^{\pm(1)}(z_1,\Pi^*)L^{\mp(2)}(z_2,\Pi^*q^{2\pi_V(h)^{(1)}})\\&\quad=L^{\mp(2)}(z_2,\Pi^*)L^{\pm(1)}(z_1,\Pi^*q^{2\pi_V(h)^{(2)}})R^{\pm*(12)}(q^{\mp c}z_1/z_2,\Pi^*).\end{aligned} \tag{2.7.6}$$

Remark 2.4 The dynamical RLL relations (2.7.5)–(2.7.6) coincide with those derived from the universal DYBE for $\mathscr{B}_{q,\lambda}(\widehat{\mathfrak{g}})$ in [76, 77].

The dynamical half currents $\widetilde{k}^{\pm}_1(z),\widetilde{k}^{\pm}_2(z),e^{\pm}(z,\Pi^*),f^{\pm}(z,\Pi^*)$ are defined similarly to (2.5.15).

$$L^{\pm}(z,\Pi^*)=\begin{pmatrix}1&f^{\pm}(z,\Pi^*)\\0&1\end{pmatrix}\begin{pmatrix}\widetilde{k}^{\pm}_1(z)&0\\0&\widetilde{k}^{\pm}_2(z)\end{pmatrix}\begin{pmatrix}1&0\\e^{\pm}(z,\Pi^*)&1\end{pmatrix}. \tag{2.7.7}$$

The relation to the previous ones is as follows.

Proposition 2.7.1

$$k_l^{\pm}(z) = \widetilde{k}_l^{\pm}(z) e^{-Q_{\bar{\epsilon}_l}} \qquad l = 1, 2, \tag{2.7.8}$$

$$e^{\pm}(z) = e^{Q_{\bar{\epsilon}_2}} e^{\pm}(z, \Pi^*) e^{-Q_{\bar{\epsilon}_1}}, \tag{2.7.9}$$

$$f^{\pm}(z) = f^{\pm}(z, \Pi^*). \tag{2.7.10}$$

Chapter 3
The H-Hopf-Algebroid Structure of $U_{q,p}(\widehat{\mathfrak{sl}}_2)_k$

In this chapter, we describe a co-algebra structure of $U_{q,p}(\widehat{\mathfrak{sl}}_2)$ given as an H-Hopf algebroid [36, 86, 97, 156]. A key idea is introducing an extended tensor product $\widetilde{\otimes}$, on which the dynamical coefficients from $\mathcal{M}_{H^*}$ get certain shifts by h and c in $\mathfrak{h}$ when they move from one tensor component to the other. These shifts produce the same effects as the dynamical shift in the DYBE and the dynamical RLL-relation. Hence the H-Hopf-algebroid structure provides a convenient co-algebra structure compatible with the dynamical shift. See Chaps. 4–8.

3.1 $\mathcal{H}$-Algebras

Let $\mathcal{A}$ be an associative algebra, $\mathcal{H}$ a finite dimensional subspace of $\mathcal{A}$, and $\mathcal{H}^*$ its dual. Denote by $\mathcal{M}_{\mathcal{H}^*}$ the field of meromorphic functions on $\mathcal{H}^*$.

Definition 3.1.1 An associative algebra $\mathcal{A}$ with 1 is said to be an $\mathcal{H}$-algebra, if it is bigraded over $\mathcal{H}^*$, $\mathcal{A} = \bigoplus_{\alpha,\beta\in\mathcal{H}^*} \mathcal{A}_{\alpha\beta}$, and equipped with two algebra embeddings $\mu_l, \mu_r : \mathcal{M}_{\mathcal{H}^*} \to \mathcal{A}_{00}$ (the left and right moment maps), such that

$$\mu_l(\widehat{f})a = a\mu_l(T_\alpha \widehat{f}), \quad \mu_r(\widehat{f})a = a\mu_r(T_\beta \widehat{f}), \qquad a \in \mathcal{A}_{\alpha\beta},\ \widehat{f} \in \mathcal{M}_{\mathcal{H}^*},$$

where T_α denotes the difference operator $(T_\alpha \widehat{f})(\lambda) = \widehat{f}(\lambda+\alpha)$ of $\mathcal{M}_{\mathcal{H}^*}$.

Let $\mathcal{A}$ and $\mathcal{B}$ be two $\mathcal{H}$-algebras.

Definition 3.1.2 An $\mathcal{H}$-algebra homomorphism from $\mathcal{A}$ to $\mathcal{B}$ is an algebra homomorphism $\pi : \mathcal{A} \to \mathcal{B}$ preserving the bigrading and the moment maps, i.e. $\pi(\mathcal{A}_{\alpha\beta}) \subseteq \mathcal{B}_{\alpha\beta}$ for all $\alpha, \beta \in \mathcal{H}^*$ and $\pi(\mu_l^{\mathcal{A}}(\widehat{f})) = \mu_l^{\mathcal{B}}(\widehat{f}), \pi(\mu_r^{\mathcal{A}}(\widehat{f})) = \mu_r^{\mathcal{B}}(\widehat{f})$.

H. Konno, *Elliptic Quantum Groups*, SpringerBriefs in Mathematical Physics 37,
https://doi.org/10.1007/978-981-15-7387-3_3

The tensor product $\mathscr{A}\widetilde{\otimes}\mathscr{B}$ is the $\mathscr{H}^*$-bigraded vector space with

$$(\mathscr{A}\widetilde{\otimes}\mathscr{B})_{\alpha\beta} = \bigoplus_{\gamma\in\mathscr{H}^*} (\mathscr{A}_{\alpha\gamma} \otimes_{\mathscr{M}_{\mathscr{H}^*}} \mathscr{B}_{\gamma\beta}),$$

where $\otimes_{\mathscr{M}_{\mathscr{H}^*}}$ denotes the usual tensor product modulo the following relation.

$$\mu_r^{\mathscr{A}}(\widehat{f})a \otimes b = a \otimes \mu_l^{\mathscr{B}}(\widehat{f})b, \qquad a \in \mathscr{A}, b \in \mathscr{B}, \widehat{f} \in \mathscr{M}_{\mathscr{H}^*}. \tag{3.1.1}$$

The tensor product $\mathscr{A}\widetilde{\otimes}\mathscr{B}$ is again an $\mathscr{H}$-algebra with the multiplication $(a\otimes b)(c\otimes d) = ac \otimes bd$ and the moment maps

$$\mu_l^{\mathscr{A}\widetilde{\otimes}\mathscr{B}} = \mu_l^{\mathscr{A}} \otimes 1, \qquad \mu_r^{\mathscr{A}\widetilde{\otimes}\mathscr{B}} = 1 \otimes \mu_r^{\mathscr{B}}.$$

Let $\mathscr{D}$ be the algebra of difference operators on $\mathscr{M}_{\mathscr{H}^*}$

$$\mathscr{D} = \{ \sum_i \widehat{f}_i T_{\beta_i} \mid \widehat{f}_i \in \mathscr{M}_{\mathscr{H}^*},\ \beta_i \in \mathscr{H}^* \}.$$

Equipped with the bigrading $\mathscr{D}_{\alpha\alpha} = \{ \widehat{f}T_{-\alpha} \mid \widehat{f} \in \mathscr{M}_{\mathscr{H}^*},\ \alpha \in \mathscr{H}^* \}$, $\mathscr{D}_{\alpha\beta} = 0$ $(\alpha \neq \beta)$ and the moment maps $\mu_l^{\mathscr{D}}, \mu_r^{\mathscr{D}} : \mathscr{M}_{\mathscr{H}^*} \to \mathscr{D}_{00}$ defined by $\mu_l^{\mathscr{D}}(\widehat{f}) = \mu_r^{\mathscr{D}}(\widehat{f}) = \widehat{f}T_0$, $\mathscr{D}$ is an $\mathscr{H}$-algebra. For any $\mathscr{H}$-algebra $\mathscr{A}$, we have the canonical isomorphism as an $\mathscr{H}$-algebra

$$\mathscr{A} \cong \mathscr{A}\widetilde{\otimes}\mathscr{D} \cong \mathscr{D}\widetilde{\otimes}\mathscr{A} \tag{3.1.2}$$

by $a \cong a\widetilde{\otimes}T_{-\beta} \cong T_{-\alpha}\widetilde{\otimes}a$ for all $a \in \mathscr{A}_{\alpha\beta}$.

Now let us consider the elliptic algebra $\mathscr{U} = U_{q,p}(\widehat{\mathfrak{sl}}_2)_k$. Let H be the same as defined in Chap. 2 and take $\mathscr{H} = H$.

Proposition 3.1.1 *The algebra $\mathscr{U}$ is an H-algebra by*

$$\mathscr{U} = \bigoplus_{\alpha,\beta\in H^*} \mathscr{U}_{\alpha,\beta} \tag{3.1.3}$$

$$\mathscr{U}_{\alpha\beta} = \left\{ a \in \mathscr{U} \,\middle|\, q^{P+h}aq^{-(P+h)} = q^{\langle\alpha,P+h\rangle}a, \quad q^P aq^{-P} = q^{\langle\beta,P\rangle}a, \quad \forall h \in \bar{\mathfrak{h}}, \forall P \in H \right\}$$

and $\mu_l, \mu_r : \mathbb{F} \to \mathscr{U}_{0,0}$ defined by [97]

$$\mu_l(\widehat{f}) = f(\Pi, p) \in \mathbb{F}, \qquad \mu_r(\widehat{f}) = f(\Pi^*, p^*) \in \mathbb{F}$$

for $\widehat{f} = f(\Pi^*, p^*) \in \mathbb{F}$.[1]

[1]Here we write the p^* (elliptic nome)-dependence explicitly in case $\widehat{f}$ is given in terms of the theta functions $\theta^*(\cdot) = \theta(\cdot, p^*)$'s.

In particular corresponding to (3.1.1), we have

$$f(\Pi^*, p^*)a\widetilde{\otimes}b = a\widetilde{\otimes}f(\Pi, p)b \qquad a, b \in \mathscr{U}. \tag{3.1.4}$$

Example 3.1 For $R^+(z,\Pi)_{ij}^{kl} \in \mathbb{F}$, we have

$$R^{*+}(z,\Pi^*)_{ij}^{kl}a\widetilde{\otimes}b = a\widetilde{\otimes}R^+(z,\Pi)_{ij}^{kl}b. \qquad \square$$

Note that from Exercise 2.3, one obtains

$$L_{ij}^+(z) \in \mathscr{U}_{-Q_{\bar{\epsilon}_i}, -Q_{\bar{\epsilon}_j}}[[z, z^{-1}]].$$

We regard $T_\alpha = e^{-Q_\alpha} \in \mathbb{C}[\mathscr{Q}_Q]$ as the shift operator

$$(T_\alpha \mu_r(\widehat{f})) = e^{-Q_\alpha} f(P, p^*)e^{Q_\alpha} = f(P + \langle Q_\alpha, P\rangle, p^*),$$
$$(T_\alpha \mu_l(\widehat{f})) = e^{-Q_\alpha} f(P + h, p)e^{Q_\alpha} = f(P + h + \langle Q_\alpha, P + h\rangle, p).$$

Then $\mathscr{D} = \{\widehat{f}e^{-Q_\alpha} \mid \widehat{f} \in \mathbb{F}, e^{-Q_\alpha} \in \mathbb{C}[\mathscr{Q}_Q]\}$ becomes the H-algebra having the property (3.1.2) for $\mathscr{A} = \mathscr{U}$.

Hereafter we abbreviate $f(\Pi, p)$ and $f(\Pi^*, p^*)$ as $f(\Pi)$ and $f^*(\Pi)$, respectively.

3.2 $\mathscr{H}$-Hopf Algebroids

We next introduce the $\mathscr{H}$-Hopf algebroid as follows [36, 86].

Definition 3.2.1 An $\mathscr{H}$-bialgebroid is an $\mathscr{H}$-algebra $\mathscr{A}$ equipped with two $\mathscr{H}$-algebra homomorphisms $\Delta : \mathscr{A} \to \mathscr{A}\widetilde{\otimes}\mathscr{A}$ (the comultiplication) and $\varepsilon : \mathscr{A} \to \mathscr{D}$ (the co-unit) such that

$$(\Delta\widetilde{\otimes}\mathrm{id}) \circ \Delta = (\mathrm{id}\widetilde{\otimes}\Delta) \circ \Delta,$$
$$(\varepsilon\widetilde{\otimes}\mathrm{id}) \circ \Delta = \mathrm{id} = (\mathrm{id}\widetilde{\otimes}\varepsilon) \circ \Delta,$$

under the identification (3.1.2).

Definition 3.2.2 An $\mathscr{H}$-Hopf algebroid is an $\mathscr{H}$-bialgebroid $\mathscr{A}$ equipped with a $\mathbb{C}$-linear map $S : \mathscr{A} \to \mathscr{A}$ (the antipode), such that

$$S(\mu_r(\widehat{f})a) = S(a)\mu_l(\widehat{f}), \quad S(a\mu_l(\widehat{f})) = \mu_r(\widehat{f})S(a), \quad \forall a \in \mathscr{A}, \widehat{f} \in \mathscr{M}_{\mathscr{H}^*},$$
$$m \circ (\mathrm{id}\widetilde{\otimes}S) \circ \Delta(a) = \mu_l(\varepsilon(a)1), \quad \forall a \in \mathscr{A},$$
$$m \circ (S\widetilde{\otimes}\mathrm{id}) \circ \Delta(a) = \mu_r(T_\alpha(\varepsilon(a)1)), \quad \forall a \in \mathscr{A}_{\alpha\beta},$$

where $m : \mathcal{A}\widetilde{\otimes}\mathcal{A} \to \mathcal{A}$ denotes the multiplication and $\varepsilon(a)1$ is the result of applying the difference operator $\varepsilon(a)$ to the constant function $1 \in \mathcal{M}_{\mathcal{H}^*}$.

Remark 3.1 ([86]) Definition 3.2.2 yields that the antipode of an $\mathcal{H}$-Hopf algebroid uniquely exists and gives the algebra antihomomorphism.

The $\mathcal{H}$-algebra $\mathcal{D}$ is an $\mathcal{H}$-Hopf algebroid with $\Delta_{\mathcal{D}} : \mathcal{D} \to \mathcal{D}\widetilde{\otimes}\mathcal{D}$, $\varepsilon_{\mathcal{D}} : \mathcal{D} \to \mathcal{D}$, $S_{\mathcal{D}} : \mathcal{D} \to \mathcal{D}$ defined by

$$\Delta_{\mathcal{D}}(\widehat{f}T_{-\alpha}) = \widehat{f}T_{-\alpha}\widetilde{\otimes}T_{-\alpha},$$
$$\varepsilon_{\mathcal{D}} = \mathrm{id}, \qquad S_{\mathcal{D}}(\widehat{f}T_{-\alpha}) = T_{\alpha}\widehat{f} = (T_{\alpha}\widehat{f})T_{\alpha}.$$

Now let us consider our H-algebra $\mathcal{U}$. We define two H-algebra homomorphisms, the co-unit $\varepsilon : \mathcal{U} \to \mathcal{D}$ and the comultiplication $\Delta : \mathcal{U} \to \mathcal{U}\widetilde{\otimes}\mathcal{U}$ by

$$\varepsilon(L^+_{ij}(z)) = \delta_{i,j}T_{\bar{\epsilon}_i} \quad (n \in \mathbb{Z}), \qquad \varepsilon(e^Q) = e^Q, \tag{3.2.1}$$

$$\varepsilon(\mu_l(\widehat{f})) = \varepsilon(\mu_r(\widehat{f})) = \widehat{f}T_0, \tag{3.2.2}$$

$$\Delta(L^+_{ij}(z)) = \sum_k L^+_{ik}(z)\widetilde{\otimes}L^+_{kj}(z), \tag{3.2.3}$$

$$\Delta(e^Q) = e^Q\widetilde{\otimes}e^Q, \qquad \Delta(\widehat{d}) = \widehat{d}\,\widetilde{\otimes}1 + 1\widetilde{\otimes}\widehat{d}, \tag{3.2.4}$$

$$\Delta(\mu_l(\widehat{f})) = \mu_l(\widehat{f})\widetilde{\otimes}1, \quad \Delta(\mu_r(\widehat{f})) = 1\widetilde{\otimes}\mu_r(\widehat{f}). \tag{3.2.5}$$

One can check that Δ preserves the defining relations of $\mathcal{U}$, which are equivalent to those in Definition C.1.1. See Appendix C and [101].

Lemma 3.2.1 *The maps ε and Δ satisfy*

$$(\Delta\widetilde{\otimes}id) \circ \Delta = (id\widetilde{\otimes}\Delta) \circ \Delta, \tag{3.2.6}$$

$$(\varepsilon\widetilde{\otimes}id) \circ \Delta = id = (id\widetilde{\otimes}\varepsilon) \circ \Delta. \tag{3.2.7}$$

Proof Straightforward. □

Hence $(\mathcal{U}, \Delta, \mathcal{M}_{H^*}, \mu_l, \mu_r, \varepsilon)$ is a H-bialgebroid.
Define further an algebra antihomomorphism (the antipode) $S : \mathcal{U} \to \mathcal{U}$ by

$$S(L^+_{ij}(z)) = (L^+(z)^{-1})_{ij}, \tag{3.2.8}$$

$$S(e^Q) = e^{-Q}, \quad S(\mu_r(\widehat{f})) = \mu_l(\widehat{f}), \quad S(\mu_l(\widehat{f})) = \mu_r(\widehat{f}). \tag{3.2.9}$$

Then S preserves the RLL-relation (2.5.18) and satisfies the antipode axioms. We hence obtain

Theorem 3.2.2 *The H-algebra $\mathscr{U}$ equipped with (Δ, ε, S) is an H-Hopf algebroid.*

Definition 3.2.3 We call the H-Hopf algebroid $(\mathscr{U}, H, \mathscr{M}_{H^*}, \mu_l, \mu_r, \Delta, \varepsilon, S)$ the elliptic quantum group $U_{q,p}(\widehat{\mathfrak{sl}}_2)_k$.

Remark 3.2 $U_{q,p}(\widehat{\mathfrak{sl}}_2)$ admits another H-algebra structure through another coproduct called the Drinfeld coproduct [76, 98].

Exercise 3.1 Show that Δ preserves the dynamical RLL-relation (2.5.18), i.e.

$$\sum_{i',j'} \Delta(R^+(z_1/z_2, \Pi)_{ij}^{i'j'})\Delta(L^+_{i'i''}(z_1))\Delta(L^+_{j'j''}(z_2))$$

$$= \sum_{k,l} \Delta(L^+_{jk}(z_2))\Delta(L^+_{il}(z_1))\Delta(R^{+*}(z_1/z_2, \Pi^*)_{kl}^{i''j''}). \qquad (3.2.10)$$

Note

$$R^{*+}(z, \Pi^* q^{2\langle Q_{\bar{\epsilon}_i}+Q_{\bar{\epsilon}_j}, P\rangle})_{ij}^{kl} = R^{*+}(z, \Pi^*)_{ij}^{kl}.$$

Remark 3.3 The coproduct for $L^+(z, \Pi^*)$ used in [45, 47] is essentially obtained from (3.2.3) via (2.7.1):

$$\Delta(L^+(z, \Pi^*)) = L^+(z, \Pi^*) \dot{\otimes} L^+(z, \Pi^* q^{2h^{(1)}}).$$

3.3 Opposite Co-algebra Structure

In Chap. 7, we use opposite comultiplication Δ'. Accordingly the H-Hopf-algebroid structure of $\mathscr{U}$ should be changed as follows.

Let $\mathscr{A}$ and $\mathscr{B}$ be two H-algebras. The tensor product $\mathscr{A}\widetilde{\otimes}\mathscr{B}$ is the H^*-bigraded vector space with

$$(\mathscr{A}\widetilde{\otimes}\mathscr{B})_{\alpha\beta} = \bigoplus_{\gamma \in H^*} (\mathscr{B}_{\gamma\beta} \otimes_{\mathbb{F}} \mathscr{A}_{\alpha\gamma}), \qquad (3.3.1)$$

where $\otimes_{\mathbb{F}}$ denotes the usual tensor product modulo the following relation.

$$\mu_l^{\mathscr{B}}(\widehat{f})b \otimes a = b \otimes \mu_r^{\mathscr{A}}(\widehat{f})a, \qquad a \in \mathscr{A},\ b \in \mathscr{B},\ \widehat{f} \in \mathbb{F}. \qquad (3.3.2)$$

The tensor product $\mathscr{A}\widetilde{\otimes}\mathscr{B}$ is again an H-algebra with the multiplication $(b\otimes a)(d\otimes c) = bd \otimes ac (a, c \in \mathscr{A}, b, d \in \mathscr{B})$ and the moment maps

$$\mu_l^{\mathscr{A}\widetilde{\otimes}\mathscr{B}} = 1 \otimes \mu_l^{\mathscr{A}}, \qquad \mu_r^{\mathscr{A}\widetilde{\otimes}\mathscr{B}} = \mu_r^{\mathscr{B}} \otimes 1. \tag{3.3.3}$$

Then the opposite comultiplication $\Delta' : \mathscr{A} \to \mathscr{A}\widetilde{\otimes}\mathscr{A}$ is defined as

$$\Delta'(L^+_{i,j}(z)) = \sum_k L^+_{k,j}(z)\widetilde{\otimes}L^+_{i,k}(z), \tag{3.3.4}$$

$$\Delta'(e^Q) = e^Q\widetilde{\otimes}e^Q, \qquad \Delta'(\widehat{d}) = \widehat{d}\,\widetilde{\otimes}1 + 1\widetilde{\otimes}\widehat{d}, \tag{3.3.5}$$

$$\Delta'(\mu_l(\widehat{f})) = 1\widetilde{\otimes}\mu_l(\widehat{f}), \quad \Delta'(\mu_r(\widehat{f})) = \mu_r(\widehat{f})\widetilde{\otimes}1. \tag{3.3.6}$$

The co-unit $\varepsilon : \mathscr{A} \to \mathscr{D}$ and the antipode $S : \mathscr{A} \to \mathscr{A}$ are the same as above.

Chapter 4
Representations of $U_{q,p}(\widehat{\mathfrak{sl}}_2)$

In this chapter we introduce a notion of dynamical representation as representation of the dynamical quantum group $U_{q,p}(\widehat{\mathfrak{sl}}_2)$. We then discuss both finite and infinite dimensional representations of $U_{q,p}(\widehat{\mathfrak{sl}}_2)$. As examples, we present the evaluation representation associated with the vector representation (Sect. 4.2) and the level-1 highest weight representations (Sect. 4.3). Most of them can be extended to $U_{q,p}(\mathfrak{g})$ for arbitrary untwisted affine Lie algebra $\mathfrak{g}$ [38, 76, 90, 95, 103].

4.1 Dynamical Representations

Let us introduce the dynamical representation of $\mathscr{U} = U_{q,p}(\widehat{\mathfrak{sl}}_2)$. Recall $H = \mathbb{C}P$, $H^* = \mathbb{C}Q$, $\mathbb{F} = \mathscr{M}_{H^*}$, and the defining relations in Definition 2.3.1.

Let $\mathscr{V}$ be a vector space over $\mathbb{F}$. We assume $\mathscr{V}$ is $\bar{\mathfrak{h}}$-diagonalizable, i.e.

$$\mathscr{V} = \bigoplus_{\mu\in\bar{\mathfrak{h}}^*} \mathscr{V}_\mu, \quad \mathscr{V}_\mu = \{v \in \mathscr{V} \mid q^h v = q^{\langle\mu,h\rangle} v \quad \forall h \in \bar{\mathfrak{h}}\}.$$

Let us define the H-algebra $\mathscr{D}_{H,\mathscr{V}}$ of the $\mathbb{C}$-linear operators on $\mathscr{V}$ by

$$\mathscr{D}_{H,\mathscr{V}} = \bigoplus_{\alpha,\beta\in H^*} (\mathscr{D}_{H,\mathscr{V}})_{\alpha\beta},$$

$$(\mathscr{D}_{H,\mathscr{V}})_{\alpha\beta} = \left\{ X \in \mathrm{End}_{\mathbb{C}}\mathscr{V} \;\middle|\; \begin{array}{c} q^{2(P+h)}Xq^{-2(P+h)} = q^{2\langle\alpha,P+h\rangle}X, \\ q^{2P}Xq^{-2P} = q^{2\langle\beta,P\rangle}X, \ \forall h \in \bar{\mathfrak{h}}, \forall P \in H \\ X\mathscr{V}_\mu \subseteq \mathscr{V}_{\mu+\alpha-\beta} \end{array} \right\},$$

$$\mu_l^{\mathscr{D}_{H,\mathscr{V}}}(\widehat{f})v = f(\Pi q^{2\langle\mu,h\rangle}, p)v, \quad \mu_r^{\mathscr{D}_{H,\mathscr{V}}}(\widehat{f})v = f(\Pi^*, p^*)v, \quad \widehat{f} \in \mathbb{F}, \ v \in \mathscr{V}_\mu.$$

H. Konno, *Elliptic Quantum Groups*, SpringerBriefs in Mathematical Physics 37,
https://doi.org/10.1007/978-981-15-7387-3_4

Definition 4.1.1 We define a dynamical representation of $\mathcal{U}$ on $\mathcal{V}$ to be an H-algebra homomorphism $\pi : \mathcal{U} \to \mathcal{D}_{H,\mathcal{V}}$. By the action π of $\mathcal{U}$ we regard $\mathcal{V}$ as a $\mathcal{U}$-module.

Definition 4.1.2 For $k \in \mathbb{C}$, we say that a $\mathcal{U}$-module has level k if $q^{c/2}$ acts as the scalar $q^{k/2}$ on it.

Let $(\pi_{\mathcal{V}}, \mathcal{V})$, $(\pi_{\mathcal{W}}, \mathcal{W})$ be two dynamical representations of $\mathcal{U}$. We define the tensor product $\mathcal{V}\widetilde{\otimes}\mathcal{W}$ by

$$\mathcal{V}\widetilde{\otimes}\mathcal{W} = \bigoplus_{\mu \in H^*} (\mathcal{V}\widetilde{\otimes}\mathcal{W})_\mu, \quad (\mathcal{V}\widetilde{\otimes}\mathcal{W})_\mu = \bigoplus_{\beta \in \bar{\mathfrak{h}}^*} \mathcal{V}_\beta \otimes_{\mathcal{M}_{H^*}} \mathcal{W}_{\mu-\beta},$$

where $\otimes_{\mathcal{M}_{H^*}}$ denotes the usual tensor product modulo the relation

$$f^*(\Pi^*)v \otimes w = v \otimes f(\Pi)w \tag{4.1.1}$$

for $v \in \mathcal{V}$, $w \in \mathcal{W}$. The action of the scalar $f^*(\Pi^*) \in \mathbb{F}$ on the tensor space $\mathcal{V}\widetilde{\otimes}\mathcal{W}$ is defined as follows.

$$f^*(\Pi^*)(v\widetilde{\otimes}w) = \Delta(\mu_r(\widehat{f}))(v\widetilde{\otimes}w) = v\widetilde{\otimes} f^*(\Pi^*)w.$$

We have a natural H-algebra embedding $\theta_{VW} : \mathcal{D}_{H,\mathcal{V}}\widetilde{\otimes}\mathcal{D}_{H,\mathcal{W}} \to \mathcal{D}_{H,\mathcal{V}\widetilde{\otimes}\mathcal{W}}$ by $X_{\mathcal{V}}\widetilde{\otimes}X_{\mathcal{W}} \in (\mathcal{D}_{H,\mathcal{V}})_{\alpha\gamma} \otimes_{\mathcal{M}_{H^*}} (\mathcal{D}_{H,\mathcal{W}})_{\gamma\beta} \mapsto X_{\mathcal{V}}\widetilde{\otimes}X_{\mathcal{W}} \in (\mathcal{D}_{H,\mathcal{V}\widetilde{\otimes}\mathcal{W}})_{\alpha\beta}$. Hence $\theta_{VW} \circ (\pi_V \otimes \pi_W) \circ \Delta : \mathcal{U} \to \mathcal{D}_{H,\mathcal{V}\widetilde{\otimes}\mathcal{W}}$ gives a dynamical representation of $\mathcal{U}$ on $\mathcal{V}\widetilde{\otimes}\mathcal{W}$.

4.2 Evaluation Representation

Let $V = \oplus_{l=1,2}\mathbb{C}v_l$ be the vector representation of $\mathfrak{sl}_2$ as in Chap. 2. Consider the space $\widehat{V}_w = V[[w, w^{-1}]] \otimes_{\mathbb{C}} \mathbb{F}$. Define $\pi_w : \mathcal{U} \to \mathcal{D}_{H,\widehat{V}_w}$ by

$$\pi_w(K^\pm) = e^{-Q/2}, \tag{4.2.1}$$

$$\pi_w(\alpha_m) = \frac{[m]_q}{m} w^m (q^{-m} E_{11} - q^m E_{22}), \tag{4.2.2}$$

$$\pi_w(e(z)) = E_{12} \frac{(pq^2; p)_\infty}{(p; p)_\infty} \delta\left(\frac{w}{z}\right) e^{-Q}, \tag{4.2.3}$$

$$\pi_w(f(z)) = E_{21} \frac{(pq^{-2}; p)_\infty}{(p; p)_\infty} \delta\left(\frac{w}{z}\right), \tag{4.2.4}$$

$$\widehat{d} = w \frac{\partial}{\partial w} \tag{4.2.5}$$

with $\pi_w(h) = E_{11} - E_{22}$, $e^{-\mathcal{Q}}v_l = v_l$ $(l = 1, 2)$. Then $(\pi_w, \widehat{V}_w)$ is a level-0 dynamical representation of $\mathscr{U}$. We call this *the evaluation representation associated with the vector representation.*

In particular, from the definition of $\psi^{\pm}(z)$, we have

$$\pi_w(\psi^+(z)) = e^{-\mathcal{Q}}\frac{\theta(q^{2\pi_w(h)}z/w)}{\theta(z/w)}, \tag{4.2.6}$$

$$\pi_w(\psi^-(z)) = e^{-\mathcal{Q}}\frac{\theta(q^{-2\pi_w(h)}w/z)}{\theta(w/z)}. \tag{4.2.7}$$

Exercise 4.1 Show that $\pi_w : \mathscr{U} \to \mathscr{D}_{H,\widehat{V}_w}$ is an H-algebra homomorphism. Hint:

$$\theta(q^{2\pi_w(h)}z) = \begin{pmatrix} \theta(q^2 z) & 0 \\ 0 & \theta(q^{-2}z) \end{pmatrix}$$

and for any $w \in \mathbb{C}^{\times}$

$$\left.\frac{\theta(wz)}{\theta(z)}\right|_{|p|<|z|<1} - \left.\frac{\theta(wz)}{\theta(z)}\right|_{1<|z|<|p^{-1}|} = -\frac{\theta(w)}{(p;p)_\infty^3}\delta(z). \tag{4.2.8}$$

To show the second formula note the well known formula

$$\left.\frac{\theta(wz)}{\theta(z)}\right|_{|p|<|z|<1} = -\frac{\theta(w)}{(p;p)_\infty^3}\sum_{n\in\mathbb{Z}}\frac{1}{1-wp^n}z^n,$$

$$\left.\frac{\theta(wz)}{\theta(z)}\right|_{1<|z|<|p^{-1}|} = \left.\frac{\theta(w^{-1}z^{-1})}{\theta(z^{-1})}\right|_{|p|<|z^{-1}|<1} = \frac{\theta(w)}{(p;p)_\infty^3}\sum_{n\in\mathbb{Z}}\frac{1}{1-w^{-1}p^n}z^{-n}. \quad \square$$

Substituting (4.2.1)–(4.2.7) into (2.5.1)–(2.5.4) with $k = 0$, one obtains

$$\pi_w(k_1^+(z)) = \rho^+(z/w)\left(E_{11} + \frac{\theta(q^{-2}z/w)}{\theta(z/w)}E_{22}\right)e^{-\mathcal{Q}_{\bar\epsilon_1}}, \tag{4.2.9}$$

$$\pi_w(k_2^+(z)) = \rho^+(z/w)\left(\frac{\theta(z/w)}{\theta(q^2 z/w)}E_{11} + E_{22}\right)e^{-\mathcal{Q}_{\bar\epsilon_2}}, \tag{4.2.10}$$

$$\pi_w(e^+(z)) = e^{\mathcal{Q}_{\bar\epsilon_2}}E_{12}\frac{\theta(\Pi^{*-1}z/w)\theta(q^2)}{\theta(z/w)\theta(\Pi^{*-1})}e^{-\mathcal{Q}_{\bar\epsilon_1}}, \tag{4.2.11}$$

$$\pi_w(f^+(z)) = E_{21}\frac{\theta(\Pi^* z/w)\theta(q^2)}{\theta(z/w)\theta(\Pi^*)}. \tag{4.2.12}$$

Note the formula

$$\oint \frac{dt}{2\pi i t} F(t)\delta(z/t) = \oint \frac{dt}{2\pi i t} F(t) \left(\frac{1}{1-z/t}\Big|_{|z/t|<1} - \frac{1}{1-z/t}\Big|_{|z/t|>1} \right)$$
$$= \oint_{|t-z|=1} \frac{dt}{2\pi i} F(t) \frac{1}{t-z} = F(z).$$

Combining all into (2.5.15), one obtains the evaluation representation of $L^+(z)$.

Proposition 4.2.1

$$\pi_w(L^+_{11}(z)) = \rho^+(z/w) \begin{pmatrix} 1 & 0 \\ 0 & b(z/w) \end{pmatrix} e^{-Q_{\bar{\epsilon}_1}}, \quad \pi_w(L^+_{12}(z)) = \rho^+(z/w) \begin{pmatrix} 0 & 0 \\ c(z/w, \Pi^*) & 0 \end{pmatrix} e^{-Q_{\bar{\epsilon}_2}},$$

$$\pi_w(L^+_{21}(z)) = \rho^+(z/w) \begin{pmatrix} 0 & \bar{c}(z/w, \Pi^*) \\ 0 & 0 \end{pmatrix} e^{-Q_{\bar{\epsilon}_1}}, \quad \pi_w(L^+_{22}(z)) = \rho^+(z/w) \begin{pmatrix} b(z/w) & 0 \\ 0 & 1 \end{pmatrix} e^{-Q_{\bar{\epsilon}_2}}.$$

Hence in terms of the dynamical L-operator in (2.7.1), *one has*

$$\pi_w(L^+_{ij}(z, \Pi^*))v_l = \sum_{l'=1,2} R^+(z/w, \Pi^*)^{jl}_{il'} v_{l'}, \tag{4.2.13}$$

where $R^+(z/w, \Pi^*)^{jl}_{il'}$ are the R-matrix given in (2.4.1).

4.3 The Level-1 Representation

Let us start by defining the level-k highest weight representation of $\mathcal{U}$.

Definition 4.3.1 For $k \in \mathbb{C}$, $\lambda, \nu \in H^*$, a $\mathcal{U}$-module $\mathcal{V}(\lambda, \nu)$ is called the level-k highest weight module with the highest weight (λ, ν), if there exists $u \in \mathcal{V}(\lambda, \nu)$ such that

$$\mathcal{V}(\lambda, \nu) = \mathcal{U} u, \qquad q^{c/2}u = q^{k/2}u,$$
$$q^{2(P+h)}u = q^{2\langle\lambda, P+h\rangle}u, \qquad q^{2P}u = q^{2\langle\nu, P\rangle}u,$$
$$\alpha_n u = f_n u = e_m u = 0 \qquad (n > 0, m \geq 0).$$

Now let Λ_0, Λ_1 be the fundamental weights of $\widehat{\mathfrak{sl}}_2$. Set $\bar{\Lambda}_0 = 0$. For $\nu \in \bar{\mathfrak{h}}^*$, consider the spaces

$$\mathcal{V}(\Lambda_a + \nu, \nu) = \mathcal{F} \otimes e^{\bar{\Lambda}_a}\mathbb{C}[\mathcal{Q}] \otimes e^{Q_\nu}\mathbb{C}[\mathcal{Q}_Q] \quad a = 0, 1, \tag{4.3.1}$$
$$\mathcal{F} = \mathbb{C}[\alpha_{-n}\ (n > 0)]. \tag{4.3.2}$$

Define the following operators on $\mathcal{V}(\Lambda_a+\nu,\nu)$.

$$\widetilde{\psi}^{\pm}(q^{-1/2}z)=\exp\left(-(q-q^{-1})\sum_{n>0}\frac{\alpha_{-n}}{1-p^n}z^n\right)\exp\left((q-q^{-1})\sum_{n>0}\frac{p^n\alpha_n}{1-p^n}z^{-n}\right)q^{-h}e^{-Q},\tag{4.3.3}$$

$$\widetilde{\psi}^{-}(q^{1/2}z)=\exp\left(-(q-q^{-1})\sum_{n>0}\frac{p^n\alpha_{-n}}{1-p^n}z^n\right)\exp\left((q-q^{-1})\sum_{n>0}\frac{\alpha_n}{1-p^n}z^{-n}\right)q^{h}e^{-Q},\tag{4.3.4}$$

$$\widetilde{e}(z)=\exp\left(\sum_{n>0}\frac{\alpha_{-n}}{[n]_q}z^n\right)\exp\left(-\sum_{n>0}\frac{\alpha_n}{[n]_q}z^{-n}\right)e^{\mathfrak{a}}z^{h+1}e^{-Q},\tag{4.3.5}$$

$$\widetilde{f}(z)=\exp\left(-\sum_{n>0}\frac{\alpha'_{-n}}{[n]_q}z^n\right)\exp\left(\sum_{n>0}\frac{\alpha'_n}{[n]_q}z^{-n}\right)e^{-\mathfrak{a}}z^{-h+1},\tag{4.3.6}$$

$$\widetilde{d}=-\sum_{m>0}\frac{m^2}{[2m]_q[m]_q}\frac{1-p^{*m}}{1-p^m}q^m\alpha_{-m}\alpha_m-\frac{h(h+2)}{4},\tag{4.3.7}$$

where we set

$$\alpha'_n=\frac{1-p^{*n}}{1-p^n}q^n\alpha_n.$$

The actions of the operators are given as follows.

$$\alpha_{-m}\cdot w=\alpha_{-m}w,\qquad \alpha_m\cdot w=\frac{[2m]_q[m]_q}{m}\frac{1-p^m}{1-p^{*m}}q^{-m}\frac{\partial}{\partial\alpha_{-m}}w\quad (m>0),\qquad w\in\mathcal{F},$$

and

$$z^{\pm h}e^{\alpha}=z^{\pm\langle\alpha,h\rangle}e^{\alpha},\qquad e^{\alpha}\in\mathbb{C}[\mathcal{Q}]\quad\text{, etc.}$$

Then we have the following statement.

Theorem 4.3.1 *$\psi^{\pm}(z)\mapsto\widetilde{\psi}^{\pm}(z)$, $e(z)\mapsto\widetilde{e}(z)$, $f(z)\mapsto\widetilde{f}(z)$, $\widehat{d}\mapsto\widetilde{d}$ with $p^*=pq^{-2}$ gives a level-1 irreducible highest weight dynamical representation of $\mathcal{U}$ on $\mathcal{V}(\Lambda_a+\nu,\nu)$ with the highest weight $(\Lambda_a+\nu,\nu)$ and the highest weight vector $1\otimes e^{\bar{\Lambda}_a}\otimes e^{Q_\nu}$.*

Proof The statement follows from the OPE formulas

$$\widetilde{e}(z_1)\widetilde{e}(z_2) = z_1^2 \frac{(z_2/z_1; p^*)_\infty (q^{-2}z_2/z_1; p^*)_\infty}{(p^*q^2 z_2/z_1; p^*)_\infty (p^* z_2/z_1; p^*)_\infty} : \widetilde{e}(z_1)\widetilde{e}(z_2) : \text{ for } |z_2/z_1| < q^2,$$

$$\widetilde{f}(z_1)\widetilde{f}(z_2) = z_1^2 \frac{(z_2/z_1; p)_\infty (q^2 z_2/z_1; p)_\infty}{(pq^{-2} z_2/z_1; p)_\infty (pz_2/z_1; p)_\infty} : \widetilde{f}(z_1)\widetilde{f}(z_2) : \text{ for } |z_2/z_1| < q^2,$$

$$\widetilde{e}(z_1)\widetilde{f}(z_2) = \frac{1}{z_1^2(1 - qz_2/z_1)(1 - q^{-1}z_2/z_1)} : \widetilde{e}(z_1)\widetilde{f}(z_2) : \text{ for } |z_2/z_1| < q,$$

$$\widetilde{f}(z_2)\widetilde{e}(z_1) = \frac{1}{z_2^2(1 - qz_1/z_2)(1 - q^{-1}z_1/z_2)} : \widetilde{e}(z_1)\widetilde{f}(z_2) : \text{ for } |z_1/z_2| < q. \qquad \square$$

Noting

$$\mathbb{C}[\mathcal{Q}] = \bigoplus_{l \in \mathbb{Z}} \mathbb{C}e^{l\alpha}, \quad \mathbb{C}[\mathcal{Q}_Q] = \bigoplus_{l \in \mathbb{Z}} \mathbb{C}e^{lQ},$$

one has

$$\mathcal{V}(\Lambda_a + \nu, \nu) = \bigoplus_{l,m \in \mathbb{Z}} \mathcal{F}_{a,\nu}(l, m),$$
$$\mathcal{F}_{a,\nu}(l, m) = \mathcal{F} \otimes e^{\bar{\Lambda}_a + l\alpha} \otimes e^{Q_\nu + mQ}. \qquad (4.3.8)$$

Chapter 5
The Vertex Operators

The vertex operators are the most important objects in representation theory of quantum groups. They are defined as certain intertwining operators of quantum group modules. In this chapter, we discuss the vertex operators of the $U_{q,p}(\widehat{\mathfrak{sl}}_2)$-modules. There are two types of them, type I and II, due to an asymmetry of the comultiplication with respect to the tensor components. By using the evaluation representation and the level-1 highest weight representation constructed in the last chapter, we solve the intertwining relations and obtain a realization of the vertex operators explicitly. Exchange relations among the vertex operators are also given. Thus obtained vertex operators become a key to the rest of this book.

5.1 The Vertex Operators of the Level-1 $U_{q,p}(\widehat{\mathfrak{sl}}_2)$-Modules

Let $\mathcal{V}(\Lambda_a + \nu, \nu)$ be the level-1 irreducible highest weight dynamical $\mathcal{U}$-module with highest weight $(\Lambda_a + \nu, \nu)$ $(a = 0, 1)$ (Sect. 4.3) and $(\pi_z, \widehat{V}_z)$ the evaluation module associated with the two-dimensional representation $V = \oplus_{\mu=1,2}\mathbb{C}v_\mu$ (Sect. 4.2).

The co-algebra structure of $\mathcal{U}$ allows the following intertwining operators.

$$\Phi^{(1-a,a)}(z) : \mathcal{V}(\Lambda_a + \nu, \nu) \to \widehat{V}_z \widetilde{\otimes} \mathcal{V}(\Lambda_{1-a} + \nu, \nu), \tag{5.1.1}$$

$$\Psi^{*(1-a,a)}(z) : \mathcal{V}(\Lambda_a + \nu, \nu) \widetilde{\otimes} \widehat{V}_z \to \mathcal{V}(\Lambda_{1-a} + \nu', \nu') \tag{5.1.2}$$

satisfying the intertwining relations

$$\Phi^{(1-a,a)}(z)\, x = \Delta(x)\Phi^{(1-a,a)}(z), \tag{5.1.3}$$

$$\Psi^{*(1-a,a)}(z)\Delta(x) = x\, \Psi^{*(1-a,a)}(z) \qquad \forall x \in \mathcal{U}. \tag{5.1.4}$$

H. Konno, *Elliptic Quantum Groups*, SpringerBriefs in Mathematical Physics 37,
https://doi.org/10.1007/978-981-15-7387-3_5

In (5.1.2), ν' denotes a shift of ν depending on the weight of vector in $\widehat{V}_z$ specified below. We call $\Phi^{(1-a,a)}(z)$, $\Psi^{*(1-a,a)}(z)$ the type I and the type II vertex operator, respectively.[1]

Let us investigate the intertwining relations in detail. Define the components of the vertex operators by

$$\Phi^{(1-a,a)}(z)\,u = \sum_{\mu=1,2} v_\mu \widetilde{\otimes} \Phi_\mu(z)u, \qquad \forall u \in \mathscr{V}(\Lambda_a+\nu,\nu), \tag{5.1.5}$$

$$\Psi^{*(1-a,a)}(z)\,(u \,\widetilde{\otimes}\, v_\mu) = \Psi^*_\mu(qz)u \qquad (\mu=1,2). \tag{5.1.6}$$

Consider the cases $x = g(\Pi), g^*(\Pi^*) \in \mathbb{F}$ in (5.1.3)–(5.1.4). From (3.1.1), (3.1.4), and (3.2.5), one obtains

$$[\Phi_\mu(z), g^*(\Pi^*)] = 0, \quad \Phi_\mu(z)g(\Pi) = g(\Pi q^{2\langle\bar{\epsilon}_\mu,h\rangle})\Phi_\mu(z), \tag{5.1.7}$$

$$[\Psi^*_\mu(z), g(\Pi)] = 0, \quad \Psi^*_\mu(z)g^*(\Pi^*) = g^*(\Pi^* q^{2\langle Q_{\bar{\epsilon}_\mu},P\rangle})\Psi^*_\mu(z). \tag{5.1.8}$$

Here one should note $hv_\mu = \langle\bar{\epsilon}_\mu, h\rangle v_\mu$ and $\langle\bar{\epsilon}_\mu, h\rangle = \langle Q_{\bar{\epsilon}_\mu}, P\rangle$. Hence $\Phi_\mu(z)$ and $\Psi^*_\mu(z)$ are linear operators

$$\Phi_\mu(z) \,:\, \mathscr{V}(\Lambda_a+\nu,\nu) \;\to\; \mathscr{V}(\Lambda_{1-a}+\nu,\nu), \tag{5.1.9}$$

$$\Psi^*_\mu(z) \,:\, \mathscr{V}(\Lambda_a+\nu,\nu) \;\to\; \mathscr{V}(\Lambda_{1-a}+\nu-\bar{\epsilon}_\mu,\nu-\bar{\epsilon}_\mu). \tag{5.1.10}$$

Note also that for the decomposition (4.3.8), one has

$$\Phi_\mu(z) \,:\, \mathscr{F}_{a,\nu}(l,m) \;\to\; \mathscr{F}_{1-a,\nu}(l+\mu-2+a,m), \tag{5.1.11}$$

$$\Psi^*_\mu(z) \,:\, \mathscr{F}_{a,\nu}(l,m) \;\to\; \mathscr{F}_{1-a,\nu-\bar{\epsilon}_\mu}(l-\mu+1+a,m). \tag{5.1.12}$$

Next let us consider the case $x = L^+(z)$.

Proposition 5.1.1 *The intertwining relations for $x = L^+(z)$ are given as follows.*

$$(id\widetilde{\otimes}\Phi^{(1-a,a)}(z_2))L^+(z_1) = R^{+(12)}(z_1/z_2,\Pi)L^{+(13)}(z_1)(id\widetilde{\otimes}\Phi^{(1-a,a)}(z_2)), \tag{5.1.13}$$

$$L^+(z_1)\Psi^{*(1-a,a)}(z_2) = \Psi^{*(1-a,a)}(z_2)L^{+(12)}(z_1)R^{*+(13)}(z_1/z_2,\Pi^* q^{-2(h^{(1)}+h^{(3)})}). \tag{5.1.14}$$

[1]Our convention of type I and type II vertex operators is different from the one in, e.g. [79], due to an oppositeness of the convention of the comultiplication Δ in (3.2.3).

These are described by the following diagrams.

$$\begin{array}{ccccc}
\widehat{V}_{z_1}\widetilde{\otimes}\mathcal{V}(\Lambda_a+\nu,\nu) & \stackrel{L^+(z_1)}{\longrightarrow} & \widehat{V}_{z_1}\widetilde{\otimes}\mathcal{V}(\Lambda_a+\nu,\nu) & \stackrel{\mathrm{id}\widetilde{\otimes}\Phi(z_2)}{\longrightarrow} & \widehat{V}_{z_1}\widetilde{\otimes}\widehat{V}_{z_2}\widetilde{\otimes}\mathcal{V}(\Lambda_{1-a}+\nu,\nu) \\
\mathrm{id}\widetilde{\otimes}\Phi(z_2)\searrow & & & & \uparrow R^+(z_1/z_2,\Pi) \\
& & \widehat{V}_{z_1}\widetilde{\otimes}\widehat{V}_{z_2}\widetilde{\otimes}\mathcal{V}(\Lambda_{1-a}+\nu,\nu) & \underset{L^{+(13)}(z_1)}{\longrightarrow} & \widehat{V}_{z_1}\widetilde{\otimes}\widehat{V}_{z_2}\widetilde{\otimes}\mathcal{V}(\Lambda_{1-a}+\nu,\nu)
\end{array}$$

$$\begin{array}{ccccc}
\widehat{V}_{z_1}\widetilde{\otimes}\mathcal{V}(\Lambda_a+\nu,\nu)\widetilde{\otimes}\widehat{V}_{z_2} & \stackrel{\mathrm{id}\widetilde{\otimes}\Psi^*(z_2)}{\longrightarrow} & \widehat{V}_{z_1}\widetilde{\otimes}\mathcal{V}(\Lambda_{1-a}+\nu',\nu') & \stackrel{L^+(z_1)}{\longrightarrow} & \widehat{V}_{z_1}\widetilde{\otimes}\mathcal{V}(\Lambda_{1-a}+\nu',\nu') \\
R^{+*(13)}(z_1/z_2,\Pi^*)\searrow & & & & \uparrow \mathit{id}\widetilde{\otimes}\Psi^*(z_2) \\
& & \widehat{V}_{z_1}\widetilde{\otimes}\mathcal{V}(\Lambda_a+\nu,\nu)\widetilde{\otimes}\widehat{V}_{z_2} & \underset{L^+(z_1)\widetilde{\otimes}\mathrm{id}}{\longrightarrow} & \widehat{V}_{z_1}\widetilde{\otimes}\mathcal{V}(\Lambda_a+\nu,\nu)\widetilde{\otimes}\widehat{V}_{z_2}
\end{array}$$

Proof

Type I:

$$\begin{aligned}
&(\mathrm{id}\,\widetilde{\otimes}\,\Phi(z_2))L^+(z_1)(v_j\,\widetilde{\otimes}\,u) \\
&\quad=\sum_i v_i\,\widetilde{\otimes}\,\Phi(z_2)L^+_{ij}(z_1)u=\sum_i\sum_\mu v_i\,\widetilde{\otimes}\,v_\mu\,\widetilde{\otimes}\,\Phi_\mu(z_2)L^+_{ij}(z_1)u.
\end{aligned}$$

On the other hand,

$$\begin{aligned}
\sum_i v_i\,\widetilde{\otimes}\,\Phi(z_2)L^+_{ij}(z_1)u&=\sum_i v_i\,\widetilde{\otimes}\,(\pi_{z_2}\otimes\mathrm{id})\Delta(L^+_{ij}(z_1))\Phi(z_2),\\
&=\sum_i v_i\,\widetilde{\otimes}\,\sum_l\pi_{z_2}(L^+_{il}(z_1))\widetilde{\otimes}L^+_{lj}(z_1)\Phi(z_2)\\
&=\sum_i v_i\,\widetilde{\otimes}\,\sum_l\sum_\rho\pi_{z_2}(L^+_{il}(z_1))v_\rho\widetilde{\otimes}L^+_{lj}(z_1)\Phi_\rho(z_2)\\
&=\sum_i v_i\,\widetilde{\otimes}\,\sum_l\sum_\rho\sum_\mu R^+(z_1/z_2,\Pi^*)^{l\rho}_{i\mu}v_\mu\widetilde{\otimes}L^+_{lj}(z_1)\Phi_\rho(z_2)\\
&=\sum_i\sum_\mu v_i\,\widetilde{\otimes}\,v_\mu\widetilde{\otimes}\sum_l\sum_\rho R^+(z_1/z_2,\Pi)^{l\rho}_{i\mu}L^+_{lj}(z_1)\Phi_\rho(z_2).
\end{aligned}$$

In the fourth equality we used (4.2.13). Hence we obtain (5.1.13) in the component form

$$\Phi_\mu(z_2)L^+_{ij}(z_1)=\sum_{\rho,l}R^{+(12)}(z_1/z_2,\Pi)^{l\rho}_{i\mu}L^+_{lj}(z_1)\Phi_\rho(z_2). \quad (5.1.15)$$

Type II:

$$L^+(z_1)(\mathrm{id}\,\widetilde{\otimes}\,\Psi^*(z_2))(v_j\,\widetilde{\otimes}\,u\,\widetilde{\otimes}\,v_\mu)$$
$$=\sum_i v_i\,\widetilde{\otimes}\,L^+_{ij}(z_1)\Psi^*(z_2)(u\,\widetilde{\otimes}\,v_\mu)=\sum_i v_i\,\widetilde{\otimes}\,L^+_{ij}(z_1)\Psi^*_\mu(qz_2)u.$$

On the other hand,

$$\sum_i v_i\,\widetilde{\otimes}\,L^+_{ij}(z_1)\Psi^*(z_2)(u\,\widetilde{\otimes}\,v_\mu)$$
$$=\sum_i v_i\,\widetilde{\otimes}\,\Psi^*(z_2)(\mathrm{id}\widetilde{\otimes}\pi_{z_2})\Delta(L^+_{ij}(z_1))(u\,\widetilde{\otimes}\,v_\mu)$$
$$=\sum_i v_i\,\widetilde{\otimes}\,\Psi^*(z_2)\left(\sum_l\sum_\rho L^+_{il}(z_1)u\,\widetilde{\otimes}R^+(z_1/z_2,\Pi^*)^{j\mu}_{l\rho}v_\rho\right)$$
$$=\sum_i v_i\,\widetilde{\otimes}\,\Psi^*(z_2)\left(\sum_l\sum_\rho R^{+*}(z_1/z_2,\Pi^*q^{-2h^{(3)}})^{j\mu}_{l\rho}L^+_{il}(z_1)u\widetilde{\otimes}v_\rho\right)$$
$$=\sum_i v_i\,\widetilde{\otimes}\,\sum_l\sum_\rho \Psi^*_\rho(qz_2)R^{+*}(z_1/z_2,\Pi^*q^{-2h^{(3)}})^{j\mu}_{l\rho}L^+_{il}(z_1)u$$
$$=\sum_i v_i\,\widetilde{\otimes}\,\sum_l\sum_\rho \Psi^*_\rho(qz_2)L^+_{il}(z_1)R^{+*}(z_1/z_2,\Pi^*q^{-2(h^{(1)}+h^{(3)})})^{j\mu}_{l\rho}u.$$

Here the third equality follows from (3.1.4), and the last equality follows from (2.5.25), i.e. $g(\Pi^*)L^+_{il}(z)=L^+_{il}(z)g(\Pi^*q^{-2\langle Q_{\bar{\epsilon}_l},P\rangle})$ and

$$\langle Q_{\bar{\epsilon}_l},P\rangle=\langle\bar{\epsilon}_l,h\rangle=h\Big|_{v_l}.$$

Hence we obtain (5.1.14) in the component form.

$$L^+_{ij}(z_1)\Psi^*_\mu(qz_2)=\sum_l\sum_\rho \Psi^*_\rho(qz_2)L^+_{il}(z_1)R^{+*}(z_1/z_2,\Pi^*q^{-2(h^{(1)}+h^{(3)})})^{j\mu}_{l\rho}. \tag{5.1.16}$$

□

Remark 5.1 The intertwining relations (5.1.13)–(5.1.14) are essentially the same as those obtained in [76] by assuming that $\mathscr{U}$, in particular $L^+(z,\Pi^*)$, follows the same co-algebra structure as the quasi-Hopf formulation $\mathscr{B}_{q,\lambda}(\widehat{\mathfrak{sl}}_2)$ [77] through the coincidence mentioned in Remark 2.4.

Remark 5.2 In the current case we take the same evaluation representations associated with V for the first and the third vector spaces in (5.1.14). In this case the dynamical shift $q^{-2(h^{(1)}+h^{(3)})}$ in $R^{+*(13)}$ has no real contribution because of the zero weight property of the R-matrix. However, in general the dynamical shift $q^{-2(h^{(1)}+h^{(3)})}$ plays a role. See, for example, the vertex operators for higher level representations in [76].

5.1.1 Type I Vertex Operator

Let us solve (5.1.7) and (5.1.15).

The first relation in (5.1.7) indicates that $\Phi_\mu(z)$ does not carry any P-weights, i.e. any dynamical charges e^{Q_α} ($\alpha \in \bar{\mathfrak{h}}^*$). Then the second relation in (5.1.7) indicates that $\Phi_\mu(z)$ carries the h-weight $-\bar{\epsilon}_\mu$, i.e. the charge $e^{-\bar{\epsilon}_\mu} \in e^{\Lambda_1}\mathbb{C}[\mathcal{Q}]$.

Next let us consider (5.1.15). Assuming that $\Phi_2(z_2)f^+(z_1)k_2^+(z_1)$ has no poles at $z_1 = pq^{-2}z_2$, one finds that the following gives the sufficient conditions.

$$\Phi_2(z_2)k_2^+(z_1) = \rho^+(z_1/z_2)k_2^+(z_1)\Phi_2(z_2), \tag{5.1.17}$$

$$\Phi_2(z_2)e(z_1) = e(z_1)\Phi_2(z_2), \tag{5.1.18}$$

$$\Phi_2(z_2)f(z_1) = \frac{\theta(q^2z_2/z_1)}{\theta(z_2/z_1)}f(z_1)\Phi_2(z_2), \tag{5.1.19}$$

$$\Phi_1(z_2) = f^+(z_2)\Phi_2(z_2), \tag{5.1.20}$$

Derivation:

- The case $\mu = i = j = 2$ yields (5.1.17).
- The case $\mu = i = 2$, $j = 1$ yields

$$\Phi_2(z_2)k_2^+(z_1)e^+(z_1) = \rho^+(z_1/z_2)k_2^+(z_1)e^+(z_1)\Phi_2(z_2).$$

Substituting (5.1.17) to this, we obtain

$$[\Phi_2(z_2), e^+(z_1)] = 0.$$

Then using (2.5.3) and (5.1.7), one obtains (5.1.18).

- The case $\mu = j = 2$, $i = 1$ yields

$$\begin{aligned}&\Phi_2(z_2)f^+(z_1)k_2^+(z_1)\\&= \rho^+(z_1/z_2)\Big(b(z_1/z_2,\Pi)f^+(z_1)k_2^+(z_1)\Phi_2(z_2) + c(z_1/z_2,\Pi)k_2^+(z_1)\Phi_1(z_2)\Big).\end{aligned} \tag{5.1.21}$$

Substituting the formula

$$\rho^+(z)=\frac{\theta(q^2z)}{\varphi(z)},\qquad \varphi(z)=q^{-3/2}\theta(q^{-2}z)\frac{\Gamma(q^{-2}z;p,q^4)\Gamma(q^6z;p,q^4)}{\Gamma(z;p,q^4)\Gamma(q^4z;p,q^4)},$$

we have

$$\begin{aligned}&\varphi(z_1/z_2)\Phi_2(z_2)f^+(z_1)k_2^+(z_1)\\&=\frac{\theta(q^2\Pi)\theta(q^{-2}\Pi)\theta(z_1/z_2)}{\theta(\Pi)^2}f^+(z_1)k_2^+(z_1)\Phi_2(z_2)+\frac{\theta(\Pi z_1/z_2)\theta(q^2)}{\theta(\Pi)}k_2^+(z_1)\Phi_1(z_2).\end{aligned}$$

Note that $\varphi(z_1/z_2)$ has a simple zero at $z_1/z_2 = pq^{-2}$. Let us suppose $\Phi_2(z_2)f^+(z_1)k_2^+(z_1)$ has no poles at $z_1/z_2=pq^{-2}$. Then setting $z_1/z_2=pq^{-2}$, we obtain

$$\begin{aligned}0=&\frac{\theta(q^2\Pi)\theta(q^{-2}\Pi)\theta(pq^{-2})}{\theta(\Pi)^2}f^+(pq^{-2}z_2)k_2^+(pq^{-2}z_2)\Phi_2(z_2)\\&+\frac{\theta(\Pi pq^{-2})\theta(q^2)}{\theta(\Pi)}k_2^+(pq^{-2}z_2)\Phi_1(z_2).\end{aligned}$$

Hence from $k_2^+(z)\Pi=q^{-2}\Pi k_2^+(z)$ we have

$$\Phi_1(z_2)=-q^{-2}\Pi\frac{\theta(q^4\Pi)}{\theta(q^2\Pi)}k_2^+(pq^{-2}z_2)^{-1}f^+(pq^{-2}z_2)k_2^+(pq^{-2}z_2)\Phi_2(z_2).$$

Simplifying the RHS by (2.5.10) one obtains (5.1.20).

- To derive (5.1.19), substitute (5.1.20) into (5.1.21). We then obtain

$$\begin{aligned}&\Phi_2(z_2)f^+(z_1)k_2^+(z_1)\\&=\rho^+(z_1/z_2)\Big(b(z_1/z_2,\Pi)f^+(z_1)k_2^+(z_1)\Phi_2(z_2)+c(z_1/z_2,\Pi)k_2^+(z_1)f^+(z_2)\Phi_2(z_2)\Big).\end{aligned}\tag{5.1.22}$$

From (2.5.10) and (2.4.11), the RHS is equal to

$$\begin{aligned}&\rho^+(z_1/z_2)\Bigg\{b(z_1/z_2,\Pi)f^+(z_1)\\&\quad+c(z_1/z_2,\Pi)\left(\frac{1}{\bar b(z_1/z_2)}f^+(z_2)-\frac{\bar c(z_1/z_2,\Pi)}{\bar b(z_1/z_2)}f^+(z_1)\right)\Bigg\}k_2^+(z_1)\Phi_2(z_2)\\&=\rho^+(z_1/z_2)\left\{\frac{\theta(q^{-2}z)}{\theta(z)}f^+(z_1)+\frac{\theta(\Pi z_1/z_2)\theta(q^2)}{\theta(\Pi)\theta(z_1/z_2)}f^+(z_2)\right\}k_2^+(z_1)\Phi_2(z_2).\end{aligned}$$

Then from (2.5.4), (2.3.1), (2.3.3), and the identity

$$\theta(q^{-2}z_1/z_2)\theta(q^{-2}\Pi z_1/w)\theta(q^{-4}\Pi)\theta(z_2/w)+\theta(q^2)\theta(q^{-4}\Pi z_1/z_2)\theta(q^{-2}\Pi z_2/w)\theta(z_1/w)$$
$$=\theta(q^2z_2)\theta(1/z_2)\theta(q^{-4}\Pi)\theta(q^{-2}\Pi),$$

this is equal to

$$\rho^+(z_1/z_2)a\oint\frac{dw}{2\pi iw}f(w)\frac{\theta(q^{-4}\Pi z_1/w)\theta(q^2)}{\theta(z_1/w)\theta(q^{-4}\Pi)}\frac{\theta(q^2z_2/w)}{\theta(z_2/w)}k_2^+(z_1)\Phi_2(z_2).$$

By using (5.1.17), (5.1.7) and comparing this with the LHS of (5.1.22), we obtain (5.1.19) as sufficient conditions.

Then we obtain the following realization of the vertex operators [76].

Theorem 5.1.2

$$\Phi_2(z)=\exp\left\{\sum_{n\neq0}\frac{\alpha'_{-n}}{[2n]_q}(qz)^n\right\}\exp\left\{-\sum_{n\neq0}\frac{\alpha'_n}{[2n]_q}(qz)^{-n}\right\}:e^{-\bar{\epsilon}_2}(-qz)^{h/2},\tag{5.1.23}$$

$$\Phi_1(z)=f^+(z)\Phi_2(z)=a\oint_C\frac{dt}{2\pi it}\Phi_2(z)f(t)\varphi(z,t;\Pi),\tag{5.1.24}$$

where

$$\varphi(z,t;\Pi)=\frac{\theta(\Pi z/t)\theta(q^2)}{\theta(q^2z/t)\theta(\Pi)}.\tag{5.1.25}$$

Proof The relations (5.1.17)–(5.1.19) follow from the following OPE formulas obtained from (2.5.2), (4.3.5)–(4.3.6), and (5.1.23).

$$\Phi_2(z_2)k_2^+(z_1)=\frac{(q^2z_1/z_2;p,q^4)^2_\infty}{(q^4z_1/z_2;p,q^4)_\infty(z_1/z_2;p,q^4)_\infty}:\Phi_2(z_2)k_2^+(z_1):,\tag{5.1.26}$$

$$k_2^+(z_1)\Phi_2(z_2)=q^{1/2}\frac{(pq^2z_2/z_1;p,q^4)^2_\infty}{(pq^4z_2/z_1;p,q^4)_\infty(pz_2/z_1;p,q^4)_\infty}:\Phi_2(z_2)k_2^+(z_1):,\tag{5.1.27}$$

$$\Phi_2(z_2)e(z_1)=(-qz_2)(1-q^{-1}z_1/z_2):\Phi_2(z_2)e(z_1):,\tag{5.1.28}$$

$$e(z_1)\Phi_2(z_2) = z_1(1 - qz_2/z_1) : \Phi_2(z_2)e(z_1) :, \tag{5.1.29}$$

$$\Phi_2(z_2)f(z_1) = (-qz_2)^{-1}\frac{(pq^{-2}z_1/z_2;p)_\infty}{(z_1/z_2;p)_\infty} : \Phi_2(z_2)f(z_1) :, \tag{5.1.30}$$

$$f(z_1)\Phi_2(z_2) = z_1^{-1}\frac{(pz_2/z_1;p)_\infty}{(q^2z_2/z_1;p)_\infty} : \Phi_2(z_2)f(z_1) : . \tag{5.1.31}$$

Note that (5.1.26) and (5.1.30) verify the assumption. The remaining conditions in (5.1.15) can also be checked by similar arguments to those in the above derivation.

□

5.1.2 *Type II Vertex Operator*

One can obtain a similar result for the type II vertex operators.

From (5.1.8) one finds that $\Psi_\mu^*(z)$ carries charges $e^{\bar{\epsilon}_\mu}$ and $e^{-Q_{\bar{\epsilon}_\mu}}$ in such a way that the combination $e^{\bar{\epsilon}_\mu}e^{-Q_{\bar{\epsilon}_\mu}}$ has weight 0 w.r.t $P+h$. On the other hand, (5.1.16) yields the following sufficient conditions.

$$k_2^+(z_1)\Psi_2^*(qz_2) = \rho^{+*}(z_1/z_2)\Psi_2^*(qz_2)k_2^+(z_1), \tag{5.1.32}$$

$$\Psi_2^*(z_2)f(z_1) = f(z_1)\Psi_2^*(z_2), \tag{5.1.33}$$

$$\Psi_2^*(z_2)e(z_1) = \frac{\theta(q^{-2}z_2/z_1)}{\theta(z_2/z_1)}e(z_1)\Psi_2^*(z_2), \tag{5.1.34}$$

$$\Psi_1^*(qz_2) = \Psi_2^*(qz_2)e^+(z_2), \tag{5.1.35}$$

under the assumption that $k_2^+(z_1)e^+(z_1)\Psi_2^*(qz_2)$ has no poles at $z_1 = p^*q^{-2}z_2$.

Then solving (5.1.32)–(5.1.35), one finds the following realization.

Theorem 5.1.3

$$\Psi_2^*(z) = \exp\left\{-\sum_{n\neq 0}\frac{\alpha_{-n}}{[2n]_q}(q^{-1}z)^n\right\}\exp\left\{\sum_{n\neq 0}\frac{\alpha_n}{[2n]_q}(q^{-1}z)^{-n}\right\}e^{\bar{\epsilon}_2}e^{Q/2}(-q^{-1}z)^{-h/2}, \tag{5.1.36}$$

$$\Psi_1^*(z) = \Psi_2^*(z)e^+(q^{-1}z), = a^*\oint_C\frac{dt}{2\pi it}\varphi^*(z,t;\Pi^{*-1})e(t)\Psi_2^*(z), \tag{5.1.37}$$

where

$$\varphi^*(z,t;\Pi^{*-1}) = \frac{\theta^*(\Pi^{*-1}q^{-2}z/t)\theta^*(q^2)}{\theta^*(z/t)\theta^*(\Pi^{*-1})}. \tag{5.1.38}$$

Proof The statement follows from the OPE formulas

$$\Psi_2^*(qz_2)k_2^+(z_1) = \frac{(q^4z_1/z_2; p^*, q^4)_\infty (z_1/z_2; p^*, q^4)_\infty}{(q^2z_1/z_2; p^*, q^4)_\infty^2} : \Psi_2^*(z_2)k_2^+(z_1) :,$$

$$k_2^+(z_1)\Psi_2^*(z_2) = q^{-1/2}\frac{(p^*q^4z_2/z_1; p^*, q^4)_\infty (p^*z_2/z_1; p^*, q^4)_\infty}{(p^*q^2z_2/z_1; p, q^4)_\infty^2} : \Psi_2^*(z_2)k_2^+(z_1) :,$$

$$\Psi_2(z_2)f(z_1) = (-q^{-1}z_2)(1 - qz_1/z_2) : \Psi_2^*(z_2)f(z_1) :,$$

$$f(z_1)\Psi_2^*(z_2) = z_1(1 - q^{-1}z_2/z_1) : \Psi_2^*(z_2)f(z_1) :,$$

$$\Psi_2(z_2)e(z_1) = -qz_2^{-1}\frac{(p^*q^2z_1/z_2; p^*)_\infty}{(z_1/z_2; p^*)_\infty} : \Psi_2^*(z_2)e(z_1) :,$$

$$e(z_1)\Psi_2^*(z_2) = z_1^{-1}\frac{(p^*z_2/z_1; p)_\infty}{(q^{-2}z_2/z_1; p^*)_\infty} : \Psi_2^*(z_2)e(z_1) : .$$

□

5.2 Exchange Relations

We finally derive the exchange relations among the vertex operators. We show that the relations are consequence of the RLL-relation (2.5.18) and the intertwining relations (5.1.13)–(5.1.14).

Lemma 5.2.1 *For level-1 vertex operators, we have*

$$\Phi_2(z_2)f^+(z_1) = k_2^+(z_2)f^+(z_1)k_2(z_2)^{-1}\Phi_2(z_2), \tag{5.2.1}$$

$$e^+(z_1)\Psi_2^*(qz_2) = \Psi_2(qz_2)k_2^+(z_2)^{-1}e^+(z_1)k_2^+(z_2). \tag{5.2.2}$$

Proof Let us consider the type I case. Applying (5.1.24) to (5.1.21), we have

$$\Phi_2(z_2)f^+(z_1)k_2^+(z_1) = \rho^+(z_1/z_2)\Big(b(z_1/z_2, \Pi)f^+(z_1)k_2^+(z_1)\Phi_2(z_2)$$
$$+c(z_1/z_2, \Pi)k_2^+(z_1)f^+(z_2)\Phi_2(z_2)\Big).$$

Then using (5.1.17), we obtain

$$\Phi_2(z_2)f^+(z_1) = \Big(b(z_1/z_2, \Pi)f^+(z_1) + c(z_1/z_2, \Pi)k_2^+(z_1)f^+(z_2)k_2^+(z_1)^{-1}\Big)\Phi_2(z_2).$$

Hence the statement follows from

$$b(z_1/z_2, \Pi)f^+(z_1) + c(z_1/z_2, \Pi)k_2^+(z_1)f^+(z_2)k_2^+(z_1)^{-1} = k_2^+(z_2)f^+(z_1)k_2(z_2)^{-1},$$

which is nothing but the $(1, 2)$, $(2, 2)$ component of the RLL-relation (3.2.10).

The type II case (5.2.2) is similar. □

Theorem 5.2.2 *The vertex operators satisfy the following exchange relations.*

$$\mathrm{id}\tilde{\otimes}\Phi(z_2)\circ\Phi(z_1) = \check{R}^{(21)}(z_1/z_2,\Pi^*)\,\mathrm{id}\tilde{\otimes}\Phi(z_1)\circ\Phi(z_2), \tag{5.2.3}$$

$$\Psi^*(z_1)\circ\Psi^*(z_2)\tilde{\otimes}\mathrm{id} = \Psi^*(z_2)\circ\Psi^*(z_1)\tilde{\otimes}\mathrm{id}\,\check{R}^{(32)}(z_1/z_2,\Pi^*), \tag{5.2.4}$$

$$\Phi(z_1)\circ\Psi^*(q^2z_2) = \left(\frac{z_1}{z_2}\right)^{1/2}\frac{\theta(qz_1/z_2,q^4)}{\theta(qz_2/z_1,q^4)}\,\mathrm{id}\tilde{\otimes}\Psi^*(q^2z_2)\circ\Phi(z_1)\tilde{\otimes}\mathrm{id}. \tag{5.2.5}$$

$$\begin{array}{ccccc}
\mathcal{V}(\Lambda_a+\nu,\nu) & \xrightarrow{\Phi(z_1)} & \widehat{V}_{z_1}\tilde{\otimes}\mathcal{V}(\Lambda_{1-a}+\nu,\nu) & \xrightarrow{\mathrm{id}\tilde{\otimes}\Phi(z_2)} & \widehat{V}_{z_1}\tilde{\otimes}\widehat{V}_{z_2}\tilde{\otimes}\mathcal{V}(\Lambda_a+\nu,\nu) \\
 & \Phi(z_2)\searrow & & & \uparrow \check{R}^{(21)}(z_1/z_2,\Pi^*) \\
 & & \widehat{V}_{z_2}\tilde{\otimes}\mathcal{V}(\Lambda_{1-a}+\nu,\nu) & \underset{\mathrm{id}\tilde{\otimes}\Phi(z_1)}{\longrightarrow} & \widehat{V}_{z_2}\tilde{\otimes}\widehat{V}_{z_1}\tilde{\otimes}\mathcal{V}(\Lambda_a+\nu,\nu)
\end{array}$$

$$\begin{array}{ccccc}
\mathcal{V}(\Lambda_a+\nu,\nu)\tilde{\otimes}\widehat{V}_{z_2}\tilde{\otimes}\widehat{V}_{z_1} & \xrightarrow{\Psi^*(z_2)\mathrm{id}} & \mathcal{V}(\Lambda_{1-a}+\nu',\nu')\tilde{\otimes}\widehat{V}_{z_1} & \xrightarrow{\Psi^*(z_1)} & \mathcal{V}(\Lambda_a+\nu'',\nu'') \\
\downarrow \check{R}^{(32)}(z_1/z_2,\Pi^*) & & & \nearrow \Psi^*(z_2) & \\
\mathcal{V}(\Lambda_a+\nu,\nu)\tilde{\otimes}\widehat{V}_{z_1}\tilde{\otimes}\widehat{V}_{z_2} & \underset{\Psi^*(z_1)\mathrm{id}}{\longrightarrow} & \mathcal{V}(\Lambda_{1-a}+\nu',\nu')\tilde{\otimes}\widehat{V}_{z_2} & &
\end{array}$$

$$\begin{array}{ccccc}
\mathcal{V}(\Lambda_a+\nu,\nu)\tilde{\otimes}\widehat{V}_{z_2} & \xrightarrow{\Psi^*(z_2)} & \mathcal{V}(\Lambda_{1-a}+\nu',\nu') & \xrightarrow{\Phi(z_1)} & \widehat{V}_{z_1}\tilde{\otimes}\mathcal{V}(\Lambda_a+\nu',\nu') \\
 & \Phi(z_2)\tilde{\otimes}\mathrm{id}\searrow & & \nearrow \mathrm{id}\tilde{\otimes}\Psi^*(z_2) & \\
 & & \widehat{V}_{z_1}\tilde{\otimes}\mathcal{V}(\Lambda_{1-a}+\nu,\nu)\tilde{\otimes}\widehat{V}_{z_2} & &
\end{array}$$

where

$$\check{R}(z,\Pi^*) = \mathsf{P}R(z,\Pi^*), \tag{5.2.6}$$

$$R(z,\Pi^*) = \mu(z)\bar{R}(z,\Pi^*), \qquad \mu(z) = z^{-1/2}\frac{\Gamma(q^4z,pz;p,q^4)}{\Gamma(q^2z,pq^2z;p,q^4)}, \tag{5.2.7}$$

$$R^*(z,\Pi^*) = -\frac{(z;p^*)_\infty(p^*q^2/z;p^*)_\infty}{(p^*q^2z;p^*)_\infty(1/z;p^*)_\infty}\mu^*(z)\bar{R}^*(z,\Pi^*). \tag{5.2.8}$$

Proof Note that in the component form, these relations are given by

$$\Phi_{\mu_2}(z_2)\Phi_{\mu_1}(z_1) = \sum_{\mu_1',\mu_2'} R(z_1/z_2,\Pi)_{\mu_1\mu_2}^{\mu_1'\mu_2'}\Phi_{\mu_1'}(z_1)\Phi_{\mu_2'}(z_2), \tag{5.2.9}$$

$$\Psi^*_{\mu_1}(z_1)\Psi^*_{\mu_2}(z_2) = \sum_{\mu_1',\mu_2'} \Psi^*_{\mu_2'}(z_2)\Psi^*_{\mu_1'}(z_1)R^*(z_1/z_2,\Pi^*)_{\mu_1'\mu_2'}^{\mu_1\mu_2}, \tag{5.2.10}$$

$$\Phi_{\mu_1}(z_1)\Psi^*_{\mu_2}(q^2 z_2) = \left(\frac{z_1}{z_2}\right)^{1/2} \frac{\theta(qz_1/z_2, q^4)}{\theta(qz_2/z_1, q^4)} \Psi^*_{\mu_2}(q^2 z_2)\Phi_{\mu_1}(z_1). \qquad (5.2.11)$$

Let us consider the type I case. The type II case is similar.

(1) $\mu_1 = \mu_2 = 2$ case: From (5.1.23), one can derive

$$\Phi_2(z_2)\Phi_2(z_1) = (-qz_2)^{1/2} \frac{(q^2 z_1/z_2, pq^2 z_1/z_2; p, q^4)_\infty}{(q^4 z_1/z_2, pz_1/z_2; p, q^4)_\infty} : \Phi_2(z_1)\Phi_2(z_2) : .$$

Hence one obtains

$$\Phi_2(z_2)\Phi_2(z_1) = \mu(z_1/z_2)\Phi_2(z_1)\Phi_2(z_2). \qquad (5.2.12)$$

(2) $\mu_1 = \mu_2 = 1$ case: From (5.1.24) the relation one needs to show is

$$f^+(z_2)\Phi_2(z_2)f^+(z_1)\Phi_2(z_1) = \mu(z_1, z_2)f^+(z_1)\Phi_2(z_1)f^+(z_2)\Phi_2(z_2).$$

From Lemma 5.2.1, one obtains

$$\begin{aligned} &f^+(z_2)k_2^+(z_2)f^+(z_1)k_2^+(z_2)^{-1}\Phi_2(z_2)\Phi_2(z_1) \\ &= \mu(z_1, z_2)f^+(z_1)k_2^+(z_1)f^+(z_2)k_2^+(z_1)^{-1}\Phi_2(z_1)\Phi_2(z_2). \end{aligned}$$

From (5.2.12), this is equivalent to

$$f^+(z_2)k_2^+(z_2)f^+(z_1)k_2^+(z_2)^{-1} = f^+(z_1)k_2^+(z_1)f^+(z_2)k_2^+(z_1)^{-1}.$$

Thanks to (2.5.7), this is nothing but the (1, 1), (2, 2) component of the RLL-relation (3.2.10).

(3) $\mu_1 = 2, \mu_2 = 1$ case: From (5.1.24) the relation one needs to show is

$$\begin{aligned} f^+(z_2)\Phi_2(z_2)\Phi_2(z_1) = \mu(z_1/z_2)\Big(&\bar{b}(z_1/z_2)\Phi_2(qz_1)f^+(z_2)\Phi_2(z_2) \\ &+\bar{c}(z_1/z_2)f^+(z_1)\Phi_2(z_1)\Phi_2(z_2)\Big). \end{aligned}$$

Applying Lemma 5.2.1, one obtains

$$\begin{aligned} f^+(z_2)\Phi_2(z_2)\Phi_2(z_1) = \mu(z_1/z_2)\Big(&\bar{b}(z_1/z_2)k_2^+(z_1)f^+(z_2)k_2^+(z_1)^{-1} \\ &+\bar{c}(z_1/z_2, \Pi)f^+(z_1)\Big)\Phi_2(z_1)\Phi_2(z_2). \end{aligned}$$

Using (5.2.12), this is equivalent to

$$f^+(z_2) = \bar{b}(z_1/z_2)k_2^+(z_1)f^+(z_2)k_2^+(z_1)^{-1} + \bar{c}(z_1/z_2, \Pi)f^+(z_1).$$

Thanks to (2.5.7), this is nothing but the (2, 1), (2, 2) component of the RLL-relation (3.2.10).

The case $\mu_1 = 1, \mu_2 = 2$ can be verified in a similar way.

The relation (5.2.11) follows from $[e(t), f(t')] = 0$ in the sense of analytic continuation and

$$\Phi_2(z_1)\Psi_2(z_2) = (-qz_1)^{-1/2}\frac{(qz_2/z_1; q^4)_\infty}{(q^{-1}z_2/z_1; q^4)_\infty} : \Phi_2(z_1)\Psi_2(z_2) :,$$

$$\Psi_2(z_2)\Phi_2(z_1) = (-q^{-1}z_2)^{-1/2}\frac{(q^5z_1/z_2; q^4)_\infty}{(q^3z_1/z_2; q^4)_\infty} : \Phi_2(z_1)\Psi_2(z_2) : . \qquad \square$$

Chapter 6
Elliptic Weight Functions

The weight functions first appeared in a construction of (q-)hypergeometric integral solutions to the (q-)KZ equations. See, for example, [115, 119, 153]. Recently it has been shown [69, 135] that they can be identified with the stable envelopes, which forms a good basis of equivariant cohomology or K-theory and plays an important role to study connections among quantum integrable systems, SUSY gauge theories, hypergeometric integrals, and geometry [117, 130]. However, until recently [99] their systematic derivations have not been written in literatures. In this chapter we present a simple derivation of them in the elliptic case. Our method is based on realizations of the vertex operators as those obtained in the last chapter and can be applied to any quantum group cases once one obtains an appropriate realization of the vertex operators. We then discuss basic properties of the elliptic weight functions such as triangular property, transition property, orthogonality, quasi-periodicity, and the shuffle product structure. The contents of this chapter are based on [99].

6.1 Combinatorial Notations

Let us consider the $U_{q,p}(\widehat{\mathfrak{sl}}_2)$ level-1 type I vertex operators $\Phi_\mu(z)$ ($\mu = 1, 2$) in Theorem 5.1.2 and their n-point composition

$$\phi(z_1, \cdots, z_n) = \Phi(z_1) \circ \cdots \circ \Phi(z_n) \ : \mathscr{V}(\Lambda_a + \nu, \nu) \to \widehat{V}_{z_n} \widetilde{\otimes} \cdots \widetilde{\otimes} \widehat{V}_{z_1} \widetilde{\otimes} \mathscr{V}(\Lambda_{a'} + \nu, \nu),$$

where $a' = a$ for even n, $1 - a$ for odd n. Its components are given by

$$\phi(z_1, \cdots, z_n) = \sum_{\mu_1, \cdots, \mu_n \in \{1,2\}} v_{\mu_n} \widetilde{\otimes} \cdots \widetilde{\otimes} v_{\mu_1} \widetilde{\otimes} \phi_{\mu_1 \cdots \mu_n}(z_1, \cdots, z_n), \quad (6.1.1)$$

$$\phi_{\mu_1 \cdots \mu_n}(z_1, \cdots, z_n) = \Phi_{\mu_1}(z_1) \cdots \Phi_{\mu_n}(z_n). \quad (6.1.2)$$

H. Konno, *Elliptic Quantum Groups*, SpringerBriefs in Mathematical Physics 37,
https://doi.org/10.1007/978-981-15-7387-3_6

For $\phi_{\mu_1\cdots\mu_n}(z_1,\cdots,z_n)$, it is convenient to introduce the following combinatorial notations. Let $[1,n]=\{1,\cdots,n\}$. Define the index set $I_l:=\{i\in[1,n]\mid \mu_i=l\}$ $(l=1,2)$ and set $\lambda_l:=|I_l|$, $\lambda:=(\lambda_1,\lambda_2)$. Then $I=(I_1,I_2)$ is a partition of $[1,n]$, i.e.

$$I_1\cup I_2=[1,n],\quad I_1\cap I_2=\emptyset.$$

We often denote thus obtained I by $I_{\mu_1\cdots\mu_n}$ and the n-point operator by $\phi_I(z_1,\cdots,z_n)$. Let $\mathbb{N}=\{m\in\mathbb{Z}|\ m\geq 0\}$. For $\lambda=(\lambda_1,\lambda_2)\in\mathbb{N}^2$ with $|\lambda|=\lambda_1+\lambda_2=n$, let $\mathscr{I}_\lambda$ be the set of all partitions $I=(I_1,I_2)$ satisfying $|I_l|=\lambda_l$ $(l=1,2)$. Note that for all $I\in\mathscr{I}_\lambda$ the n-point operators $\phi_I(z_1,\cdots,z_n)$ have the same h-weight $-\sum_{j=1}^n\bar{\epsilon}_{\mu_j}$. We call $\sum_{j=1}^n\bar{\epsilon}_{\mu_j}$ the weight associated with λ. We also set $I_1=\{i_1<\cdots<i_{\lambda_1}\}$.

Let us consider an explicit realization of $\phi_I(z_1,\cdots,z_n)$ applying Theorem 5.1.2. Remember that for $i\in I_1$, one has $\Phi_{\mu_i}(z_i)=\Phi_1(z_i)$, which is screened by the half current $f^+(z)$, whereas for $j\in I_2$, one has a bare vertex $\Phi_{\mu_j}(z_j)=\Phi_2(z_j)$. In a construction of weight functions, it is also important to label the arguments of the elliptic currents appearing in $\phi_I(z_1,\cdots,z_n)$ systematically. For $i_a\in I_1$, we assign the argument t_a to the elliptic current f attached to the i_a-th vertex operator.

Example 6.1 Let us consider the case $n=5$, $\lambda=(3,2)$. For example, the 5-point operator

$$\phi_{21121}(z_1,z_2.z_3,z_4,z_5)=\Phi_2(z_1)\Phi_1(z_2)\Phi_1(z_3)\Phi_2(z_4)\Phi_1(z_5)\qquad(6.1.3)$$

gives a partition $I=(I_1=\{2,3,5\},I_2=\{1,4\})$. Hence $i_1=2,i_2=3,i_3=5$. Then from Theorem 5.1.2 we have the following realization of $\Phi_1(z_i)$ $(i=2,3,5)$ in (6.1.3).

$$\Phi_1(z_2)=a\oint_C\frac{dt_1}{2\pi i t_1}\Phi_2(z_2)f(t_1)\varphi(z_2,t_1;\Pi),$$

$$\Phi_1(z_3)=a\oint_C\frac{dt_2}{2\pi i t_2}\Phi_2(z_3)f(t_2)\varphi(z_3,t_2;\Pi),$$

$$\Phi_1(z_5)=a\oint_C\frac{dt_3}{2\pi i t_3}\Phi_2(z_5)f(t_3)\varphi(z_5,t_3;\Pi).\qquad\square$$

6.2 Derivation of the Weight Function

Substituting the expressions of the vertex operators (5.1.23)–(5.1.24) into the n-point operator $\phi_{\mu_1\cdots\mu_n}(z_1,\cdots,z_n)$, we obtain a multiple contour integral of a product of operators $\Phi_2(z_i)$, $f(t_a)$'s and the integration kernels $\varphi(z_i,t_a;\Pi)$'s. We then divide the integrand into two parts, the operator part $\widetilde{\Phi}(t,z)$ and the kinematical factor part $\omega_{\mu_1\cdots\mu_n}(t,z,\Pi)$. The operator part consists of a normal ordered product of the bare vertex operators $\Phi_2(z)$'s, a normal ordered product of the elliptic currents $f(t_a)$ $(a = 1, \cdots, \lambda_1)$ and the symmetric part of the OPE coefficients $< \Phi_2(z_k)\Phi_2(z_l) >^{Sym}$'s and $< f(t_a)f(t_b) >^{Sym}$'s given below. The kinematical factor part consists of all $\varphi(z, t_a; \Pi)$'s, all factors arising from the exchange relations between $\Phi_2(z)$'s and $f(t_a)$'s as well as all the non-symmetric part of the OPE coefficients.

The procedure consists of the following 4 steps.

1. Move all $\varphi(z_i, t_a; \Pi)$'s to the right end. Then the dynamical parameters in φ get shift following the exchange relation

$$\Pi\Phi_\mu(z) = \Phi_\mu(z)\Pi q^{-2\langle\bar{\epsilon}_\mu, h\rangle}.$$

2. Move all the elliptic currents $f(t_a)$'s to the right of all the bare vertex operators $\Phi_2(z)$'s and put $\Phi_2(z)$'s and $f(t_a)$'s in the definite ordering $\Phi_2(z_1)\cdots\Phi_2(z_n)$ $f(t_1)\cdots f(t_{\lambda_1})$. In this process one gets appropriate factors by the exchange relations (5.1.19).
3. Take normal ordering of all $\Phi_2(z)$'s and $f(t_a)$'s, respectively. Then one gets appropriate factors following the rule

$$\Phi_2(z_1)\cdots\Phi_2(z_n) =: \Phi_2(z_1)\cdots\Phi_2(z_n) : \prod_{1\le k<l\le n} < \Phi_2(z_k)\Phi_2(z_l) >,$$

$$f(t_1)\cdots f(t_{\lambda_1}) =: f(t_1)\cdots f(t_{\lambda_1}) : \prod_{1\le a<b\le\lambda_1} < f(t_a)f(t_b) > .$$

Here the OPE coefficients are given by

$$\begin{aligned} < \Phi_2(z_k)\Phi_2(z_l) > &= (-z_k)^{1/2}\frac{\{q^2 z_l/z_k\}\{pq^2 z_l/z_k\}}{\{pz_l/z_k\}\{q^4 z_l/z_k\}} \\ &= (-z_k)^{1/2}\frac{\Gamma(q^2 z_l/z_k; p, q^4)}{\Gamma(q^4 z_l/z_k; p, q^4)} < \Phi_2(z_k)\Phi_2(z_l) >^{Sym}, \\ < f(t_a)f(t_b) > &= t_a^2\frac{(q^2 t_b/t_a; p)_\infty(t_b/t_a; p)_\infty}{(pt_b/t_a; p)_\infty(pq^{-2}t_b/t_a; p)_\infty} \\ &= \frac{\theta(q^2 t_b/t_a)}{\theta(t_b/t_a)} < f(t_a)f(t_b) >^{Sym} \end{aligned}$$

with the symmetric parts

$$\begin{aligned} < \Phi_2(z_k)\Phi_2(z_l) >^{Sym} &= \frac{(q^2 z_k/z_l, q^2 z_l/z_k; p, q^4)_\infty}{(q^4 z_k/z_l, q^4 z_l/z_k; p, q^4)_\infty}, \\ < f(t_a)f(t_b) >^{Sym} &= -qt_a t_b\frac{(t_a/t_b; p)_\infty(t_b/t_a; p)_\infty}{(pq^{-2}t_a/t_b; p)_\infty(pq^{-2}t_b/t_a; p)_\infty}. \end{aligned}$$

4. Symmetrize the integrant w.r.t. the integration variables $t_1, \cdots, t_{\lambda_1}$. We denote this procedure by Sym_t.

Now let us apply this procedure to (6.1.2). One obtains the following.

$$
\begin{aligned}
&\phi_{\mu_1\cdots\mu_n}(z_1,\cdots,z_n)\\
&=\Phi_2(z_1)\cdots\Phi_1(z_{i_1})\cdots\Phi_1(z_{i_2})\cdots\Phi_1(z_{i_{\lambda_1}})\cdots\Phi_2(z_n)\\
&=\oint_{C^{\lambda_1}}\underline{dt_1}\cdots\underline{dt_{\lambda_1}}\,\Phi_2(z_1)\cdots\Phi_2(z_n)f(t_1)\cdots f(t_{\lambda_1})\prod_{a=1}^{\lambda_1}\prod_{l=i_a+1}^{n}\frac{\theta(z_l/t_a)}{\theta(q^2z_l/t_a)}\\
&\qquad\times\prod_{a=1}^{\lambda_1}\varphi(z_{i_a},t_a;\Pi q^{-2\sum_{k=i_a+1}^{n}\langle\bar{\epsilon}_{\mu_k},h\rangle})\\
&=\oint_{C^{\lambda_1}}\underline{dt_1}\cdots\underline{dt_{\lambda_1}}\,:\Phi_2(z_1)\cdots\Phi_2(z_n)::f(t_1)\cdots f(t_{\lambda_1}):\prod_{a=1}^{\lambda_1}\prod_{l=i_a+1}^{n}\frac{\theta(z_l/t_a)}{\theta(q^2z_l/t_a)}\\
&\qquad\times\prod_{1\le k<l\le n}<\Phi_2(z_k)\Phi_2(z_l)>\prod_{1\le a<b\le\lambda_1}<f(t_a)f(t_b)>\\
&\qquad\times\prod_{a=1}^{\lambda_1}\varphi(z_{i_a},t_a;\Pi q^{-2C(i_a)}),
\end{aligned}
$$

where we set

$$
\underline{dt_k}=a\frac{dt_k}{2\pi i t_k},
$$

$$
C(i_a)=\sum_{k>i_a}^{n}\langle\bar{\epsilon}_{\mu_k},h\rangle.
$$

Hence we obtain

Theorem 6.2.1

$$
\phi_{\mu_1\cdots\mu_n}(z_1,\cdots,z_n)=\oint_{C^{\lambda_1}}\underline{dt}\,\widetilde{\Phi}(t,z)\omega_{\mu_1\cdots\mu_n}(t,z,\Pi),
$$

where we set $t=(t_1,\cdots,t_{\lambda_1})$, $z=(z_1,\cdots,z_n)$ *and*

$$
\begin{aligned}
\widetilde{\Phi}(t,z)&=:\Phi_2(z_1)\cdots\Phi_2(z_n)::f(t_1)\cdots f(t_{\lambda_1}):\\
&\times\prod_{1\le k<l\le n}<\Phi_2(z_k)\Phi_2(z_l)>^{Sym}\prod_{1\le a<b\le\lambda_1}<f(t_a)f(t_b)>^{Sym},
\end{aligned}
\quad (6.2.1)
$$

$$\omega_{\mu_1 \cdots \mu_n}(t, z, \Pi) = \mu^+(z) \widetilde{W}_I(t, z, \Pi), \tag{6.2.2}$$

$$\mu^+(z) = \prod_{1 \le k < l \le n} (-z_k)^{1/2} \frac{\Gamma(q^2 z_l / z_k; p, q^4)}{\Gamma(q^4 z_l / z_k; p, q^4)}, \tag{6.2.3}$$

$$\widetilde{W}_I(t, z, \Pi) = \mathrm{Sym}_t \prod_{a=1}^{\lambda_1} \left[\frac{\theta(\Pi q^{-2C(i_a)} z_{i_a}/t_a)\theta(q^2)}{\theta(q^2 z_{i_a}/t_a)\theta(\Pi q^{-2C(i_a)})} \prod_{b>i_a}^{n} \frac{\theta(z_b/t_a)}{\theta(q^2 z_b/t_a)} \right]$$
$$\times \prod_{1 \le a < b \le \lambda_1} \frac{\theta(q^2 t_b/t_a)}{\theta(t_b/t_a)}. \tag{6.2.4}$$

Note that $\widetilde{\Phi}(t, z)$ is an operator valued symmetric function in $z_1, \cdots, z_n$ as well as in $t_1, \cdots, t_{\lambda_1}$. The dynamical shift $C(i_a)$ has the following combinatorial expression.

Proposition 6.2.2 *For* $a = 1, \cdots, \lambda_1$,

$$C(i_a) = \sum_{k>i_a}^{n} \left(\delta_{\mu_k,1} - \delta_{\mu_k,2} \right) = 2(\lambda_1 - a) - n + i_a.$$

Let us consider the function $\widetilde{W}_I(t, z, \Pi)$. Multiplying $\widetilde{W}_I$ by

$$H_\lambda(t, z) := \prod_{a=1}^{\lambda_1} \prod_{b=1}^{n} \theta(q^2 z_b/t_a), \tag{6.2.5}$$

one obtains an entire function in t's and z's

$$W_I(t, z, \Pi) := \widetilde{W}_I(t, z, \Pi) H_\lambda(t, z) = \mathrm{Sym}_t U_I(t, z, \Pi), \tag{6.2.6}$$

$$U_I(t, z, \Pi) = \prod_{a=1}^{\lambda_1} \left[\frac{\theta(\Pi q^{-2C(i_a)} z_{i_a}/t_a)\theta(q^2)}{\theta(\Pi q^{-2C(i_a)})} \prod_{b>i_a} \theta(z_b/t_a) \prod_{b<i_a} \theta(q^2 z_b/t_a) \right]$$
$$\times \prod_{1 \le a < b \le \lambda_1} \frac{\theta(q^2 t_b/t_a)}{\theta(t_b/t_a)}. \tag{6.2.7}$$

Furthermore, in order to compare with the stable envelopes, it is convenient to consider the following expression. See Sect. 9.4.

$$\mathscr{W}_I(t, z, \Pi) = \frac{W_I(t, z, \Pi)}{E_\lambda(t)} = \mathrm{Sym}_t\, \mathscr{U}_I(t, z, \Pi), \tag{6.2.8}$$

$$\mathscr{U}_I(t, z, \Pi) = \frac{\prod_{a=1}^{\lambda_1} u_I(t_a, z, \Pi q^{-2C(i_a)})}{\prod_{1 \le a < b \le \lambda_1} \theta(t_a/t_b)\theta(q^{-2} t_b/t_a)}, \tag{6.2.9}$$

where

$$E_\lambda(t) = \prod_{a,b=1}^{\lambda_1} \theta(q^2 t_b/t_a), \tag{6.2.10}$$

$$u_I(t_a, z, \Pi) = \prod_{\substack{b=1\\ b>i_a}}^{n} \theta(z_b/t_a) \cdot \frac{\theta(\Pi z_{i_a}/t_a)}{\theta(\Pi)} \cdot \prod_{\substack{b=1\\ b<i_a}}^{n} \theta(q^2 z_b/t_a). \tag{6.2.11}$$

We call $\mathscr{W}_I(t, z, \Pi)$ the elliptic weight function of type $\mathfrak{sl}_2$.

6.3 Properties of the Elliptic Weight Functions

In this section we discuss some basic properties of the elliptic weight function $\mathscr{W}_I(t, z, \Pi)$.

6.3.1 Triangular Property

For $I, J \in \mathscr{I}_\lambda$, let $I_1 = \{i_1 < \cdots < i_{\lambda_1}\}$ and $J_1 = \{j_1 < \cdots < j_{\lambda_1}\}$. Define a partial ordering $\leqslant$ by

$$I \leqslant J \Leftrightarrow i_a \leq j_a \qquad \forall a.$$

Let us denote by $t = z_I$ the specialization $t_a = z_{i_a}$ $(a = 1, \cdots, \lambda_1)$. The weight function has the following triangular property.

Proposition 6.3.1 *For $I, J \in \mathscr{I}_\lambda$,*

(1) $\widetilde{W}_J(z_I, z, \Pi) = 0$ *unless* $I \leqslant J$.
(2)

$$\mathscr{W}_I(z_I, z, \Pi) = \prod_{i\in I_1}\left(\prod_{\substack{b\in I_2\\ i<b}} \theta(z_b/z_i) \prod_{\substack{b\in I_2\\ i>b}} \theta(q^2 z_b/z_i)\right).$$

Proof (1) follows from the factor $\displaystyle\prod_{a=1}^{\lambda_1}\prod_{b>i_a} \theta(z_b/t_a)$. □

For $\sigma \in \mathfrak{S}_n$, let us denote $\sigma^{-1}(I) = I_{\mu_{\sigma(1)}\cdots\mu_{\sigma(n)}}$ and $\sigma(z) = (z_{\sigma(1)}, \cdots, z_{\sigma(n)})$. Following [135], set $\mathscr{W}_{\sigma,I}(t, z, \Pi) = \mathscr{W}_{\sigma^{-1}(I)}(t, \sigma(z), \Pi)$ and $\mathscr{W}_{\mathrm{id},I}(t, z\Pi) = \mathscr{W}_I(t, z, \Pi)$. Let us consider the matrix $\widehat{W}_\sigma(z, \Pi)$, whose (I, J) element is given by $\mathscr{W}_{\sigma,J}(z_I, z, \Pi)$ $(I, J \in \mathscr{I}_\lambda)$. We put the matrix elements in the decreasing order with respect to $\leqslant$. Then Proposition 6.3.1 yields that the matrix $\widehat{W}_{\rm id}(z, \Pi)$ is lower triangular, whereas $\widehat{W}_{\sigma_0}(z, \Pi)$ for the longest element $\sigma_0 \in \mathfrak{S}_n$ is upper triangular. In particular, for generic z_a $(a = 1, \cdots, n)$, $\widehat{W}_\sigma(z, \Pi)$ is invertible.

6.3.2 Transition Property

Proposition 6.3.2 *Let* $I = I_{\mu_1\cdots\mu_i\mu_{i+1}\cdots\mu_n} \in \mathscr{I}_\lambda$.

$$\begin{aligned}&\mathscr{W}_{I_{\dots\ \mu_{i+1}\mu_i \dots}}(t, \cdots, z_{i+1}, z_i, \cdots, \Pi)\\&= \sum_{\mu_i', \mu_{i+1}'} \overline{R}(z_i/z_{i+1}, \Pi q^{-2\sum_{j=i}^n \langle \bar\epsilon_{\mu_j}, h\rangle})_{\mu_i\mu_{i+1}}^{\mu_i'\mu_{i+1}'} \mathscr{W}_{I_{\dots\ \mu_i'\mu_{i+1}' \dots}}(t, \cdots, z_i, z_{i+1}, \cdots, \Pi).\end{aligned} \tag{6.3.1}$$

Proof From Theorem 6.2.1 we have

$$\phi_{\cdots\mu_{i+1}\mu_i\cdots}(\cdots, z_{i+1}, z_i, \cdots) = \oint_{C^M} \underline{dt}\, \widetilde{\Phi}(t, z)\, \omega_{\dots\mu_{i+1}\mu_i\cdots}(t, \cdots, z_{i+1}, z_i, \cdots; \Pi),$$

where we used the symmetry property of $\widetilde{\Phi}(t, z)$ under any permutations of $z_1, \cdots, z_n$. Using the exchange relation (5.2.9) in the left-hand side, we obtain

$$\begin{aligned}&\phi_{\dots\mu_{i+1}\mu_i\dots}(\cdots, z_{i+1}, z_i, \cdots)\\&= \sum_{\mu_i', \mu_{i+1}'} R(z_i/z_{i+1}, \Pi q^{2\sum_{j=1}^{i-1} \langle \bar\epsilon_{\mu_j}, h\rangle})_{\mu_i\mu_{i+1}}^{\mu_i'\mu_{i+1}'} \phi_{\dots\mu_i'\mu_{i+1}'\dots}(\cdots, z_i, z_{i+1}, \cdots)\\&= \oint_{C^M} \underline{dt}\, \widetilde{\Phi}(t, z) \sum_{\mu_i', \mu_{i+1}'} R(z_i/z_{i+1}, \Pi q^{-2\sum_{j=i}^n \langle \bar\epsilon_{\mu_j}, h\rangle})_{\mu_i\mu_{i+1}}^{\mu_i'\mu_{i+1}'}\\&\qquad \times \omega_{\dots\mu_i'\mu_{i+1}'\dots}(t, \cdots, z_i, z_{i+1}, \cdots; \Pi).\end{aligned}$$

Note that $\mu(z)$ (5.2.7) in the R matrix is related to $\mu^+(z)$ (6.2.3) by

$$\mu(z_i/z_{i+1}) = \mu^+(\cdots, z_{i+1}, z_i, \cdots)/\mu^+(\cdots, z_i, z_{i+1}, \cdots).$$

Comparing the integrand we obtain the desired relation. □

Let $s_i = (i\ i+1) \in \mathfrak{S}_n (i = 1, \cdots, n-1)$ denote the adjacent transpositions. Let us set

$$\mathscr{R}^{(s_i,\mathrm{id})}(z,\Pi)_I^I = \overline{R}(z_i/z_{i+1}, \Pi q^{2\sum_{j=1}^{i-1}\langle\bar{\epsilon}_{\mu_j},h\rangle})_{\mu_i\mu_{i+1}}^{\mu_i\mu_{i+1}},$$
$$\mathscr{R}^{(s_i,\mathrm{id})}(z,\Pi)_I^{s_i(I)} = \overline{R}(z_i/z_{i+1}, \Pi q^{2\sum_{j=i}^{i-1}\langle\bar{\epsilon}_{\mu_j},h\rangle})_{\mu_i\mu_{i+1}}^{\mu_{i+1}\mu_i}.$$

Then one can rewrite (6.3.1) as

$$\mathscr{W}_{s_i,I}(t,z,\Pi)$$
$$= \begin{cases} \mathscr{W}_I(t,z,\Pi) & \text{if } s_i(I) = I \\ \mathscr{R}^{(s_i,\mathrm{id})}(z,\Pi q^{-2\sum_{j=1}^{n}\langle\bar{\epsilon}_{\mu_j},h\rangle})_I^I\, \mathscr{W}_{\mathrm{id},I}(t,z,\Pi) & \\ +\mathscr{R}^{(s_i,\mathrm{id})}(z,\Pi q^{-2\sum_{j=1}^{n}\langle\bar{\epsilon}_{\mu_j},h\rangle})_I^{s_i(I)}\, \mathscr{W}_{\mathrm{id},s_i(I)}(t,z,\Pi) & \text{if } s_i(I) \neq I \end{cases}.$$

In general, let us consider the n-point operator

$$\phi_{\sigma^{-1}(I)}(\sigma(z)) = \Phi_{\mu_{\sigma(1)}}(z_{\sigma(1)}) \cdots \Phi_{\mu_{\sigma(n)}}(z_{\sigma(n)}).$$

By using the exchange relation (5.2.9) repeatedly one obtains

$$\phi_{\sigma^{-1}(I)}(\sigma(z)) = \sum_{I'} \mathscr{R}^{(\sigma,\sigma')}(z,\Pi)_I^{I'} \phi_{\sigma'^{-1}(I')}(\sigma'(z)),$$

where $\mathscr{R}^{(\sigma,\sigma')}(z,\Pi)_I^{I'}$ denotes a coefficient given by a certain sum of products of the R matrix elements in (2.4.4). Then by the same argument as in the proof of Proposition 6.3.2 we obtain

$$\mathscr{W}_{\sigma,I}(t,z,\Pi) = \sum_{I'} \mathscr{R}^{(\sigma,\sigma')}(z,\Pi q^{-2\sum_{j=1}^{n}\langle\bar{\epsilon}_{\mu_j},h\rangle})_I^{I'} \mathscr{W}_{\sigma',I'}(t,z,\Pi). \quad (6.3.2)$$

Let us define the matrix $\mathscr{R}^{(\sigma,\sigma')}(z,\Pi)$ by

$$\mathscr{R}^{(\sigma,\sigma')}(z,\Pi) = \left(\mathscr{R}^{(\sigma,\sigma')}(z,\Pi)_I^J\right)_{I,J\in\mathscr{I}_\lambda}.$$

Then one can rewrite (6.3.2)

$$\widehat{W}_\sigma(z,\Pi) = \widehat{W}_{\sigma'}(z,\Pi)\, {}^t\mathscr{R}^{(\sigma,\sigma')}(z,\Pi q^{-2\sum_{j=1}^{n}\langle\bar{\epsilon}_{\mu_j},h\rangle}). \quad (6.3.3)$$

or

$$\widehat{W}_{\sigma'}(z,\Pi)^{-1}\, \widehat{W}_\sigma(z,\Pi) = {}^t\mathscr{R}^{(\sigma,\sigma')}(z,\Pi q^{-2\sum_{j=1}^{n}\langle\bar{\epsilon}_{\mu_j},h\rangle}). \quad (6.3.4)$$

By taking the transposition and the shift $\Pi \mapsto \Pi q^{2\sum_{j=1}^{n}\langle \bar{\epsilon}_{\mu_j}, h\rangle}$, one obtains

$$ {}^t\widehat{W}_\sigma(z, \Pi q^{2\sum_{j=1}^{n}\langle \bar{\epsilon}_{\mu_j}, h\rangle}) \left({}^t\widehat{W}_{\sigma'}(z, \Pi q^{2\sum_{j=1}^{n}\langle \bar{\epsilon}_{\mu_j}, h\rangle})\right)^{-1} = \mathscr{R}^{(\sigma,\sigma')}(z, \Pi). \quad (6.3.5) $$

In addition, one has

Proposition 6.3.3

$$ {}^t\mathscr{R}^{(\sigma,\sigma')}(z, \Pi q^{-2\sum_{j=1}^{n}\langle \bar{\epsilon}_{\mu_j}, h\rangle}) = \mathscr{R}^{(\sigma,\sigma')}(z, \Pi^{-1}). \quad (6.3.6) $$

Proof It is enough to show the case that $\mathscr{R}^{(\sigma,\sigma')}(z, \Pi)$ is given by

$$ R^{(23)}(z_2/z_3, \Pi)R^{(13)}(z_1/z_3, \Pi q^{2h^{(2)}})R^{(12)}(z_1/z_2, \Pi). $$

The desired equality follows from the properties of the R matrix (2.4.2) such as ${}^tR(z, \Pi) = R(z, \Pi^{-1})$ and $R(z, \Pi q^{2(h^{(1)}+h^{(2)})}) = R(z, \Pi)$ as well as the dynamical Yang–Baxter equation (2.4.6). □

Example 6.2 Let us consider the case $n = 3, \lambda = (2, 1)$. The weight associated with λ is $\sum_{j=1}^{3}\langle \bar{\epsilon}_{\mu_j}, h\rangle = 1$ for any $I_{\mu_1\mu_2\mu_3} \in \mathscr{I}_\lambda$. From Proposition 6.3.2, one obtains for $\sigma_0 \in \mathfrak{S}_3$

$$ \begin{aligned} &\mathscr{W}_{\sigma_0, I_{211}}(t, z, \Pi) \\ &\quad = \overline{R}(z_{13}, q^2\Pi)_{21}^{12}\mathscr{W}_{I_{112}}(t, z, \Pi) + \overline{R}(z_{13}, q^2\Pi)_{21}^{21}\overline{R}(z_{12}, \Pi)_{21}^{12}\mathscr{W}_{I_{121}}(t, z, \Pi) \\ &\quad\quad +\overline{R}(z_{13}, q^2\Pi)_{21}^{21}\overline{R}(z_{12}, \Pi)_{21}^{21}\mathscr{W}_{I_{211}}(t, z, \Pi), \end{aligned} \quad (6.3.7) $$

$$ \begin{aligned} &\mathscr{W}_{\sigma_0, I_{121}}(t, z, \Pi) \\ &\quad = \overline{R}(z_{23}, \Pi)_{21}^{12}\overline{R}(z_{13}, q^2\Pi)_{12}^{12}\mathscr{W}_{I_{112}}(t, z, \Pi) \\ &\quad\quad + \left(\overline{R}(z_{23}, \Pi)_{21}^{12}\overline{R}(z_{13}, q^2\Pi)_{12}^{21}\overline{R}(z_{12}, \Pi)_{21}^{12} + \overline{R}(z_{23}, \Pi)_{21}^{21}\overline{R}(z_{12}, \Pi)_{12}^{12}\right)\mathscr{W}_{I_{121}}(t, z, \Pi) \\ &\quad\quad +\overline{R}(z_{13}, \Pi)_{21}^{21}\overline{R}(z_{12}, q^2\Pi)_{12}^{21}\mathscr{W}_{I_{211}}(t, z, \Pi) \end{aligned} \quad (6.3.8) $$

$$ \begin{aligned} &\mathscr{W}_{\sigma_0, I_{112}}(t, z, \Pi) \\ &\quad = \overline{R}(z_{23}, \Pi)_{12}^{12}\overline{R}(z_{13}, q^2\Pi)_{12}^{12}\mathscr{W}_{I_{112}}(t, z, \Pi) + \overline{R}(z_{23}, q^2\Pi)_{12}^{21}\overline{R}(z_{13}, \Pi)_{12}^{12}\mathscr{W}_{I_{121}}(t, z, \Pi) \\ &\quad\quad +\overline{R}(z_{13}, \Pi)_{12}^{21}\mathscr{W}_{I_{211}}(t, z, \Pi). \end{aligned} \quad (6.3.9) $$

Here we set $z_{ij} = z_i/z_j$ and used the following identities.

$$ \begin{aligned} &\overline{R}(z_{23}, \Pi)_{21}^{12}\overline{R}(z_{13}, q^2\Pi)_{12}^{21}\overline{R}(z_{12}, \Pi)_{21}^{21} + \overline{R}(z_{23}, \Pi)_{21}^{21}\overline{R}(z_{12}, \Pi)_{12}^{21} = \overline{R}(z_{13}, \Pi)_{21}^{21}\overline{R}(z_{12}, q^2\Pi)_{12}^{21}, \\ &\overline{R}(z_{23}, \Pi)_{12}^{12}\overline{R}(z_{13}, q^2\Pi)_{12}^{21}\overline{R}(z_{12}, \Pi)_{21}^{12} + \overline{R}(z_{23}, \Pi)_{12}^{21}\overline{R}(z_{12}, \Pi)_{12}^{12} = \overline{R}(z_{23}, q^2\Pi)_{12}^{21}\overline{R}(z_{13}, \Pi)_{12}^{12}, \\ &\overline{R}(z_{23}, \Pi)_{12}^{12}\overline{R}(z_{13}, q^2\Pi)_{12}^{21}\overline{R}(z_{12}, \Pi)_{21}^{21} + \overline{R}(z_{23}, \Pi)_{12}^{21}\overline{R}(z_{12}, \Pi)_{12}^{21} = \overline{R}(z_{13}, \Pi)_{12}^{21}. \end{aligned} $$

Hence

$$\mathscr{R}^{(\sigma_0,\mathrm{id})}(z,\Pi)_{I_{211}}^{I_{211}} = \overline{R}(z_{13},q^2\Pi)_{21}^{21}\overline{R}(z_{12},\Pi)_{21}^{21},$$

$$\mathscr{R}^{(\sigma_0,\mathrm{id})}(z,\Pi)_{I_{211}}^{I_{121}} = \overline{R}(z_{13},q^2\Pi)_{21}^{21}\overline{R}(z_{12},\Pi)_{21}^{12},$$

$$\mathscr{R}^{(\sigma_0,\mathrm{id})}(z,\Pi)_{I_{211}}^{I_{112}} = \overline{R}(z_{13},q^2\Pi)_{21}^{12},$$

$$\mathscr{R}^{(\sigma_0,\mathrm{id})}(z,\Pi)_{I_{121}}^{I_{211}} = \overline{R}(z_{12},q^2\Pi)_{12}^{21}\overline{R}(z_{13},\Pi)_{21}^{21},$$

$$\mathscr{R}^{(\sigma_0,\mathrm{id})}(z,\Pi)_{I_{121}}^{I_{121}} = \overline{R}(z_{23},\Pi)_{21}^{12}\overline{R}(z_{13},q^2\Pi)_{12}^{21}\overline{R}(z_{12},\Pi)_{21}^{12} + \overline{R}(z_{23},\Pi)_{21}^{21}\overline{R}(z_{12},\Pi)_{12}^{12},$$

$$\mathscr{R}^{(\sigma_0,\mathrm{id})}(z,\Pi)_{I_{121}}^{I_{112}} = \overline{R}(z_{23},\Pi)_{21}^{12}\overline{R}(z_{13},q^2\Pi)_{12}^{12},$$

$$\mathscr{R}^{(\sigma_0,\mathrm{id})}(z,\Pi)_{I_{112}}^{I_{211}} = \overline{R}(z_{13},\Pi)_{12}^{21},$$

$$\mathscr{R}^{(\sigma_0,\mathrm{id})}(z,\Pi)_{I_{112}}^{I_{121}} = \overline{R}(z_{23},q^2\Pi)_{12}^{21}\overline{R}(z_{13},\Pi)_{12}^{12},$$

$$\mathscr{R}^{(\sigma_0,\mathrm{id})}(z,\Pi)_{I_{112}}^{I_{112}} = \overline{R}(z_{23},\Pi)_{12}^{12}\overline{R}(z_{13},q^2\Pi)_{12}^{12}.$$

Then it is not so hard to check Proposition 6.3.3. For example, one obtains

$$\mathscr{R}^{(\sigma_0,\mathrm{id})}(z,q^{-2}\Pi)_{I_{211}}^{I_{211}} = \bar{b}(z_{13})\bar{b}(z_{12}) = \mathscr{R}^{(\sigma_0,\mathrm{id})}(z,\Pi^{-1})_{I_{211}}^{I_{211}},$$

$$\mathscr{R}^{(\sigma_0,\mathrm{id})}(z,q^{-2}\Pi)_{I_{121}}^{I_{211}} = c(z_{12},\Pi)\bar{b}(z_{13}) = \mathscr{R}^{(\sigma_0,\mathrm{id})}(z,\Pi^{-1})_{I_{211}}^{I_{121}},$$

$$\mathscr{R}^{(\sigma_0,\mathrm{id})}(z,q^{-2}\Pi)_{I_{112}}^{I_{211}} = c(z_{13},q^{-2}\Pi) = \mathscr{R}^{(\sigma_0,\mathrm{id})}(z,\Pi^{-1})_{I_{211}}^{I_{112}}.$$

The most non-trivial one is $\mathscr{R}^{(\sigma_0,\mathrm{id})}(z,q^{-2}\Pi)_{I_{121}}^{I_{121}} = \mathscr{R}^{(\sigma_0,\mathrm{id})}(z,\Pi^{-1})_{I_{121}}^{I_{121}}$. The proof of this identity is left as an exercise for the reader. □

6.3.3 Orthogonality

This is an elliptic and dynamical analogue of the same property given in [135].

Proposition 6.3.4 *For $J, K \in \mathscr{I}_\lambda$,*

$$\sum_{I\in\mathscr{I}_\lambda} \frac{\mathscr{W}_J(z_I,z,\Pi^{-1}q^{2\sum_{j=1}^n\langle\bar{\epsilon}_{\mu_j},h\rangle})\mathscr{W}_{\sigma_0(K)}(z_I,\sigma_0(z),\Pi)}{Q(z_i)R(z_I)} = \delta_{J,K},$$

where $\sum_{j=1}^n \bar{\epsilon}_{\mu_j}$ is the weight associated with λ (Sec.6.1), and

$$Q(z_I) = \prod_{a\in I_1}\prod_{b\in I_2}\theta(q^2z_b/z_a),$$

$$R(z_I) = \prod_{a\in I_1}\prod_{b\in I_2}\theta(z_b/z_a).$$

Proof From (6.3.4), (6.3.5), and (6.3.6) with $\sigma = \sigma_0$ and $\sigma' = \mathrm{id}$, we have

$$\widehat{W}_{\mathrm{id}}(z, \Pi)^{-1}\widehat{W}_{\sigma_0}(z, \Pi) = {}^t\widehat{W}_{\sigma_0}(z, \Pi^{-1}q^{2\sum_{j=1}^n \langle \bar{\epsilon}_{\mu_j}, h\rangle}) \left({}^t\widehat{W}_{\mathrm{id}}(z, \Pi^{-1}q^{2\sum_{j=1}^n \langle \bar{\epsilon}_{\mu_j}, h\rangle})\right)^{-1}.$$

Hence

$$\widehat{W}_{\mathrm{id}}(z, \Pi)\, {}^t\widehat{W}_{\sigma_0}(z, \Pi^{-1}q^{2\sum_{j=1}^n \langle \bar{\epsilon}_{\mu_j}, h\rangle}) = \widehat{W}_{\sigma_0}(z, \Pi)\, {}^t\widehat{W}_{\mathrm{id}}(z, \Pi^{-1}q^{2\sum_{j=1}^n \langle \bar{\epsilon}_{\mu_j}, h\rangle}).$$

Since the LHS is a lower triangular matrix and the RHS is an upper triangular matrix, this must be a diagonal matrix. Let us denote it by S. Its diagonal entries are obtained from Proposition 6.3.1 (2) as

$$S_{II} = \mathscr{W}_I(z_I, z, \Pi)\mathscr{W}_{\sigma_0(I)}(z_I, \sigma_0(z), \Pi^{-1}q^{2\sum_{j=1}^n \langle \bar{\epsilon}_{\mu_j}, h\rangle}) = Q(z_I)R(z_I).$$

We then obtain

$${}^t\widehat{W}_{\mathrm{id}}(z, \Pi^{-1}q^{2\sum_{j=1}^n \langle \bar{\epsilon}_{\mu_j}, h\rangle})S^{-1}\widehat{W}_{\sigma_0}(z, \Pi) = \mathrm{id}.$$

Taking the (J, K) component of this relation, we obtain the desired result. □

6.3.4 *Quasi-periodicity*

From (1.3.1) and Proposition 6.2.2 we obtain the following statement.

Proposition 6.3.5 *For $I \in \mathscr{I}_\lambda$, the weight function $\mathscr{W}_I(t, z, \Pi)$ has the following quasi-periodicity.*

$$\mathscr{W}_I(\cdots, e^{2\pi i}t_a, \cdots, z, \Pi) = (-)^{n-2\lambda_1+2}\mathscr{W}_I(t, z, \Pi),$$

$$\mathscr{W}_I(\cdots, pt_a, \cdots, z, \Pi) = (-p^{-1/2})^{n-2\lambda_1+2}q^{2(n-\lambda_1)}t_a^{-(n-2\lambda_1)}\Pi \prod_{b=1}^{\lambda_1} t_b^{-2} \prod_{l=1}^{n} z_l\, \mathscr{W}_I(t, z, \Pi).$$

Furthermore, noting

$$H_\lambda(\cdots, e^{2\pi i}t_a, \cdots, z) = (-1)^n H_\lambda(t, z),$$

$$H_\lambda(\cdots, pt_a, \cdots, z) = (-p^{-1/2})^n q^{2n} t_a^{-n} \prod_{l=1}^{n} z_l\, H_\lambda(t, z),$$

$$E_\lambda(\cdots, e^{2\pi i}t_a, \cdots) = (-1)^{2\lambda_1-2}E_\lambda(t),$$

$$E_\lambda(\cdots, pt_a, \cdots) = (-p^{-1/2})^{2\lambda_1-2}t_a^{-2\lambda_1} \prod_{b=1}^{\lambda_1} t_b^2\, E_\lambda(t),$$

one finds

Corollary 6.3.6

$$\widetilde{W}_I(\cdots, e^{2\pi i}t_a, \cdots, z, \Pi) = \widetilde{W}_I(t, z, \Pi),$$
$$\widetilde{W}_I(\cdots, pt_a, \cdots, z, \Pi) = q^{-2\lambda_1}\Pi\ \widetilde{W}_I(t, z, \Pi).$$

For $\lambda = (\lambda_1, \lambda_2) \in \mathbb{N}^2$, $|\lambda| = n$, let $z^{(n)} = (z_1, \cdots, z_n) \in (\mathbb{C}^\times)^n$.

Definition 6.3.1 We define $\mathscr{M}_\lambda^{(n)}(z^{(n)}, \Pi_\lambda)$ to be the space of meromorphic functions $F(t; z, \Pi)$ of λ_1 variables $t = (t_1, \cdots, t_{\lambda_1})$ such that

(1) $F(t; z, \Pi)$ is symmetric in $t_1, \cdots, t_{\lambda_1}$.
(2) $F(t; z, \Pi)$ has the quasi-periodicity

$$F(\cdots, pt_a, \cdots; z, \Pi) = q^{-2\lambda_1}\Pi\ F(t; z, \Pi).$$

Let us consider the subspace $\mathscr{M}_\lambda^{+(n)}(z^{(n)}, \Pi) := \mathrm{Span}_\mathbb{C}\{\ \widetilde{W}_I(t, z, \Pi)\ (I \in \mathscr{I}_\lambda)\ \}$ of $\mathscr{M}_\lambda^{(n)}(z, \Pi)$. From Proposition 6.3.1, one obtains

Proposition 6.3.7 $\dim_\mathbb{C} \mathscr{M}_\lambda^{+(n)}(z^{(n)}, \Pi) = \dfrac{n!}{\lambda_1!(n-\lambda_1)!}$.

Remark 6.1 Let $x = {}^t(x_1, \cdots, x_{\lambda_1}) \in \mathbb{C}^{\lambda_1}$. From Proposition 6.3.5 one can deduce a symmetric integral $\lambda_1 \times \lambda_1$ matrix N and a vector $\xi \in (\mathbb{C}/\mathbb{Z})^{\lambda_1}$, which yield the following quadratic form $N(x) = {}^t xNx$ and the linear form $\xi(x) = {}^t x\xi$.

$$N(x) = n\sum_{a=1}^{\lambda_1} x_a^2,$$

$$\xi(x) = -\sum_{a=1}^{\lambda_1} x_a((P+h)+n-\lambda_1) - \sum_{a=1}^{\lambda_1}\sum_{k=1}^{n} x_a u_k.$$

Then by Appell–Humbert theorem [108], a pair (N, ξ) characterizes a line bundle $\mathscr{L}(N, \xi) : (\mathbb{C}^{\lambda_1} \times \mathbb{C})/\Lambda^{\lambda_1} \to \mathbb{C}^{\lambda_1}$, where $\Lambda = \mathbb{Z} + \mathbb{Z}\tau$, with action

$$\omega \cdot (x, \eta) = (x + \omega, e_\omega(x)\eta), \qquad \omega \in \Lambda^{\lambda_1},\ x \in \mathbb{C}^{\lambda_1},\ \eta \in \mathbb{C},$$

and cocycle

$$e_{n+m\tau}(x) = (-1)^{{}^t nNn}(-e^{i\pi\tau})^{{}^t mNm} e^{2\pi i\ {}^t m(Nx+\xi)}, \qquad n, m \in \mathbb{Z}^{\lambda_1}.$$

Moreover $\Theta_\lambda^+(z, \Pi_\lambda) := \mathrm{Span}_\mathbb{C}\{\ W_I(t, z, \Pi)\ (I \in \mathscr{I}_\lambda)\ \}$ is a space of sections of $\mathscr{L}(N, \xi)$. See, for example, [51].

6.3.5 Shuffle Algebra Structure

Consider a graded $\mathbb{C}$-vector space

$$\mathcal{M}(z,\Pi)=\bigoplus_{n\in\mathbb{N}}\bigoplus_{\substack{\lambda\in\mathbb{N}^2\\|\lambda|=n}}\mathcal{M}^{(n)}_{\lambda}(z^{(n)},\Pi)$$

with $\mathcal{M}^{(0)}_{(0,0)}(z^{(0)},\Pi)=\mathbb{C}1$.

Definition 6.3.2 For $F(t;z^{(m)},\Pi)\in\mathcal{M}^{(m)}_{\lambda}(z^{(m)},\Pi)$, $G(t';z'^{(n)},\Pi)\in{\mathcal{M}^{(n)}}_{\lambda'}(z'^{(n)},\Pi)$, we define the bilinear product $\star$ on $\mathcal{M}(z,\Pi)$ by

$$\begin{aligned}&(F\star G)(t_1,\cdots,t_{\lambda_1+\lambda'_1},\Pi)\\&:=\frac{1}{\lambda_1!\lambda'_1!}\mathrm{Sym}_t\left[F(t,z,\Pi q^{-2\sum_{j=1}^{n}\langle\bar{\epsilon}_{\mu'_j},h\rangle})\,G(t',z',\Pi)\,\Xi(t,t',z,z')\right],\end{aligned}\qquad(6.3.10)$$

where $I'=I'_{\mu'_1\cdots\mu'_n}$ and

$$\Xi(t,t',z,z')=\prod_{a=1}^{\lambda_1}\left(\prod_{b=1}^{n}\frac{\theta(z'_b/t_a)}{\theta(q^2z'_b/t_a)}\prod_{c=1}^{\lambda'_1}\frac{\theta(q^2t'_c/t_a)}{\theta(t'_c/t_a)}\right).$$

In the LHS of (6.3.10), we set $t_{\lambda_1+a}:={t'}_a$ $(a=1,\cdots,{\lambda'}_1)$, $z_{m+k}:=z'_k$ $(k=1,\cdots,n)$.

This endows $\mathcal{M}(z,\Pi)$ with a structure of an associative unital algebra with the unit 1. In [99], a $\mathfrak{sl}_N$ version of the $\star$-product is given.

Let us consider the subspace of $\mathcal{M}(z,\Pi)$.

$$\mathcal{M}^{+}(z,\Pi)=\bigoplus_{n\in\mathbb{N}}\bigoplus_{\substack{\lambda\in\mathbb{N}^2\\|\lambda|=n}}\mathcal{M}^{+(n)}_{\lambda}(z^{(n)},\Pi_\lambda).$$

From (6.2.6)–(6.2.7) it turns out that all the elements in $\mathcal{M}^{+}(z,\Pi)$ satisfy the following pole and wheel conditions. For $F(t;z,\Pi)\in\mathcal{M}^{+(n)}_{\lambda}(z^{(n)},\Pi)$,

(1) there exists an entire function $f(t;z,\Pi)\in\Theta^{+}_{\lambda}(z^{(n)},\Pi)$ such that

$$F(t;z,\Pi)=\frac{f(t;z,\Pi)}{H_\lambda(t,z)},$$

where $H_\lambda(t,z)$ is given in (6.2.5).

(2) $f(t;z,\Pi)=0$ once $t_a/z_c=q^2$ and $z_c/t_b=1$ for some a,b,c, where $a,b=1,\cdots,\lambda_1$, $c=1,\cdots,n$.

Proposition 6.3.8 *The subspace $\mathscr{M}^+(z,\Pi) \subset \mathscr{M}(z,\Pi)$ is $\star$-closed.*

Proof Let $I = I_{\mu_1\cdots\mu_m} \in \mathscr{I}_\lambda$ and $I' = I_{\mu'_1\cdots\mu'_n} \in \mathscr{I}_{\lambda'}$ and define $I + I' = (I_1 \cup I'_1, I_2 \cup I'_2) \in \mathscr{I}_{(\lambda_1+\lambda'_1,\lambda_2+\lambda'_2)}$. The statement follows from a simple fact that the composition of the m-point and the n-point operators $\phi_I(z^{(m)})$ and $\phi_{I'}(z^{(n)})$ of the form in Theorem 6.2.1 with $\omega_I(t, z^{(m)}, \Pi) = \mu^+(z^{(m)})\widetilde{W}_I(t, z^{(m)}, \Pi)$ and $\omega_{I'}(t', z^{(n)}, \Pi) = \mu^+(z^{(n)})\widetilde{W}_I(t', z^{(n)}, \Pi)$, respectively, is the $m+n$-point operator $\phi_{I\cup I'}(z^{(m)} \cup z^{(n)})$ of the same form with $\omega_{I+I'}(t \cup t', z^{(m)} \cup z^{(n)}, \Pi) = \mu^+(z^{(m)} \cup z^{(n)})\widetilde{W}_{I+I'}(t \cup t', z^{(m)} \cup z^{(n)}, \Pi)$. In the composition one needs to arrange the order of operators in the integral following the four step procedure in the derivation of Theorem 6.2.1. Then one gets the dynamical shift in $\widetilde{W}_I(t', z^{(n)}, \Pi)$ by $q^{-2(\sum_{j=1}^n \langle \bar{\epsilon}_{\mu'_j}, h\rangle)}$ and the extra factor $\Xi(t, t', z, z')$. □

Remark 6.2 The subalgebra $(\mathscr{M}^+(z,\Pi), \star)$ is an $\mathfrak{sl}_2$-type elliptic and dynamical analogue of the shuffle algebra studied in [44, 128], where the $A^{(1)}_{N-1}$-type was discussed. $(\mathscr{M}^+(z,\Pi), \star)$ is also similar to the elliptic algebra discussed in [43].

Chapter 7
Tensor Product Representation

In this chapter, we discuss a tensor product representation of the n evaluation representations constructed in Chap. 4. We introduce the Gelfand-Tsetlin basis of the tensor product space and construct an action of the elliptic currents and the half currents of $U_{q,p}(\widehat{\mathfrak{sl}}_2)$ on it. Remarkably, the change of basis matrix from the standard basis to the Gelfand-Tsetlin basis is given by a specialization of the elliptic weight functions. The resultant action is expressed in a perfectly combinatorial way in terms of the partitions of $[1, n]$. In Chap. 9 we discuss a geometric interpretation of it.

7.1 Tensor Product of the Evaluation Representations

Let $(\pi_z, \widehat{V}_z)$ be the evaluation representation discussed in Chap. 4. The action of $\mathcal{U} = U_{q,p}(\widehat{\mathfrak{sl}}_2)$ on the tensor product space $\widehat{V}_w \widetilde{\otimes} \widehat{V}_{z_1} \widetilde{\otimes} \cdots \widetilde{\otimes} \widehat{V}_{z_n}$ is given by the co-algebra structure presented in Chap. 3. Let Δ' be the opposite comultiplication (3.3.4). Noting the coassociativity define

$$\Delta'^{(1)}(x) = \Delta'(x),$$
$$\Delta'^{(m)}(x) = (\Delta' \widetilde{\otimes} \underbrace{\mathrm{id} \widetilde{\otimes} \cdots \widetilde{\otimes} \mathrm{id}}_{m-1}) \circ \Delta'^{(m-1)}(x) \qquad m = 2, 3, \cdots, \qquad \forall x \in \mathcal{U}.$$

Proposition 7.1.1 *$L^{\pm}(w)$ acts on $\widehat{V}_w \widetilde{\otimes} \widehat{V}_{z_1} \widetilde{\otimes} \cdots \widetilde{\otimes} \widehat{V}_{z_n}$ by*

$$(\pi_{z_1} \widetilde{\otimes} \cdots \widetilde{\otimes} \pi_{z_n}) \Delta'^{(n-1)}(L^{\pm}(w))$$
$$= \bar{R}^{(0n)}(w/z_n, \Pi^* q^{2\sum_{j=1}^{n-1} h^{(j)}}) \overline{R}^{(0n-1)}(w/z_{n-1}, \Pi^* q^{2\sum_{j=1}^{n-2} h^{(j)}}) \cdots \bar{R}^{(01)}(w/z_1, \Pi^*).$$

Proof It is enough to show the $n = 2$ case.

H. Konno, *Elliptic Quantum Groups*, SpringerBriefs in Mathematical Physics 37,
https://doi.org/10.1007/978-981-15-7387-3_7

$$(\pi_{z_1}\widetilde{\otimes}\pi_{z_2})\Delta'(L^{\pm}_{ij}(w))v_\mu\widetilde{\otimes}v_\nu = \sum_k \pi_{z_1}(L^{\pm}_{kj}(w))v_\mu\widetilde{\otimes}\pi_{z_2}(L^{\pm}_{ik}(w))v_\nu$$

$$= \sum_{k,\mu',\nu'} \overline{R}(w/z_1,\Pi^*)^{j\mu}_{k\mu'}v_{\mu'}\widetilde{\otimes}\overline{R}(w/z_2,\Pi^*)^{k\nu}_{i\nu'}v_{\nu'}$$

$$= \sum_{\mu',\nu'}\sum_k \overline{R}(w/z_2,\Pi^* q^{2h^{(1)}})^{k\nu}_{i\nu'}\overline{R}(w/z_1,\Pi^*)^{j\mu}_{k\mu'}v_{\mu'}\widetilde{\otimes}v_{\nu'}.$$

The third equality follows from (3.3.2). □

It is also useful to write down the comultiplication formula of the dynamical L-operator, which is equivalent to Proposition 7.1.1.

Proposition 7.1.2 *The dynamical L-operator $L^+(w,\Pi^*)$ acts on $\widehat{V}_w \otimes \widehat{V}_{z_1} \otimes \cdots \otimes \widehat{V}_{z_n}$, where $\otimes$ denotes the usual tensor product, by*

$$(\pi_{z_1}\otimes\cdots\otimes\pi_{z_n})\Delta'^{(n-1)}(L^+_{ij}(w,\Pi^*))$$
$$= \sum_{k_1,\cdots,k_{n-1}\in\{1,2\}} L^+_{k_1 j}(w/z_1,\Pi^*)\otimes L^+_{k_2k_1}(w/z_1,\Pi^* q^{2h^{(1)}})\otimes\cdots\otimes L^+_{ik_{n-1}}(w/z_n,\Pi^* q^{2\sum_{j=1}^{n-1}h^{(j)}}).$$

Note that this is equivalent to consider the action

$$(\pi_{z_1}\otimes\cdots\otimes\pi_{z_n})\Delta'^{(n-1)}(L^+_{ij}(w))$$
$$= \sum_{k_1,\cdots,k_{n-1}\in\{1,2\}} L^+_{k_1 j}(w/z_1)\widetilde{\otimes}L^+_{k_2k_1}(w/z_1)\widetilde{\otimes}\cdots\widetilde{\otimes}L^+_{ik_{n-1}}(w/z_n) \quad (7.1.1)$$

on $\widehat{V}_w\widetilde{\otimes}\widehat{V}_{z_1}\widetilde{\otimes}\cdots\widetilde{\otimes}\widehat{V}_{z_n}$ with the rule (3.3.2).

7.2 The Gelfand-Tsetlin Basis

Definition 7.2.1 At the level 0 the Heisenberg subalgebra generated by α_m ($m \in \mathbb{Z}\backslash\{0\}$) becomes the maximal commutative subalgebra of $\mathscr{U}$. We call this the Gelfand-Tsetlin (GT) subalgebra $\mathfrak{G}$.

Definition 7.2.2 The Gelfand-Tsetlin basis is a basis of the level-0 representation of $\mathscr{U}$ consisting of the simultaneous eigenvectors of the GT subalgebra $\mathfrak{G}$.

The GT basis first appeared in the work of Fedor Smirnov on form factors of massive integrable QFT [150], where it was used for the construction of solutions to the rational q-KZ equation. It was rediscovered by Maillet and others [114] on algebraic Bethe ansatz, under the name of F-basis. Extension to the elliptic dynamical version was worked out by [1]. We here follow yet another construction formulated in [135] and construct the GT basis of $\mathscr{U}$-module $\widehat{V}_{z_1}\widetilde{\otimes}\cdots\widetilde{\otimes}\widehat{V}_{z_n}$ explicitly.

Firstly we realize $\mathfrak{S}_n$ in terms of the elliptic dynamical R matrix in (2.4.2). Let us define $\widetilde{S}_i(P)$ by

$$\widetilde{S}_i(\Pi^*) := \mathsf{P}^{(ii+1)}\overline{R}^{(ii+1)}(z_{i+1}/z_i, \Pi^* q^{2\sum_{j=1}^{i-1} h^{(j)}})s_i^z,$$

where

$$\mathsf{P} : v\widetilde{\otimes}w \mapsto w\widetilde{\otimes}v, \qquad s_i^z f(\cdots, z_i, z_{i+1}, \cdots) = f(\cdots, z_{i+1}, z_i, \cdots).$$

Then by using the DYBE (2.4.6) and the unitarity relation (2.4.7) one can show the following.

Proposition 7.2.1

$$\begin{aligned}
&\widetilde{S}_i(\Pi^*)\widetilde{S}_{i+1}(\Pi^*)\widetilde{S}_i(\Pi^*) = \widetilde{S}_{i+1}(\Pi^*)\widetilde{S}_i(\Pi^*)\widetilde{S}_{i+1}(\Pi^*),\\
&\widetilde{S}_i(\Pi^*)\widetilde{S}_j(\Pi^*) = \widetilde{S}_j(\Pi^*)\widetilde{S}_i(\Pi^*) \qquad\qquad (|i-j|>1)\\
&\widetilde{S}_i(\Pi^*)^2 = 1.
\end{aligned}$$

For a vector $v_{\mu_1}\widetilde{\otimes}\cdots\widetilde{\otimes}v_{\mu_n}$, $\mu_1, \cdots, \mu_n \in \{1, 2\}$, we define the partition $I = (I_1, I_2)$ of $[1, n]$ in the same was as in Sect. 6.1. For $\lambda \in \mathbb{N}^2$, $|\lambda| = n$, $I = I_{\mu_1\cdots\mu_n} \in \mathscr{I}_\lambda$, we set

$$v_I = v_{\mu_1\cdots\mu_n} := v_{\mu_1}\widetilde{\otimes}\cdots\widetilde{\otimes}v_{\mu_n}.$$

Define $\{\xi_I\}_{I\in\mathscr{I}_\lambda}$ by

$$\xi_{I^{max}} := v_{I^{max}}, \qquad \xi_{s_i(I)} := \widetilde{S}_i(\Pi^*)\xi_I, \tag{7.2.1}$$

where

$$I^{max} = I_{\underbrace{2\cdots2}_{\lambda_2}\underbrace{1\cdots1}_{\lambda_1}}$$

denotes the maximal partition in $\mathscr{I}_\lambda$ w.r.t the partial ordering $\leqslant$ in Sect. 6.3.1. Then $\{\xi_I\}_{I\in\mathscr{I}_\lambda}$ gives the GT basis. This is due to the fact that $L^+_{22}(z) = k^+_2(z)$ is diagonal on $v_{I^{max}}$ and the following statement.

Proposition 7.2.2

$$\widetilde{S}_i(\Pi^*)\Delta'^{(n-1)}(L^\pm(w)) = \Delta'^{(n-1)}(L^\pm(w))\widetilde{S}_i(\Pi^* q^{2h^{(0)}}).$$

Proof Use the DYBE. □

Let us consider the change of basis matrix $\widehat{X} = (X_{IJ}(z, \Pi^*))_{I,J\in\mathscr{I}_\lambda}$:

$$\xi_I = \sum_{J\in\mathscr{I}_\lambda} X_{IJ}(z, \Pi^*)v_J. \tag{7.2.2}$$

Putting the matrix elements in the decreasing order $I^{max} \geqslant \cdots \geqslant I^{min}$, $\widehat{X}$ is by construction a lower triangular matrix. It is then remarkable that one can realize $\widehat{X}$ explicitly in terms of the elliptic weight functions. This turns out to be a key to obtain a geometric interpretation of the results in this chapter. See Sect. 9.6.

Theorem 7.2.3

$$X_{IJ}(z, \Pi^*) = \widetilde{W}_J(z_I, z, \Pi^* q^{2\sum_{j=1}^{n}\langle\bar{\epsilon}_{\mu_j}, h\rangle}). \tag{7.2.3}$$

The proof is given in Appendix D.

Example 7.1 Let us consider the case $N = 2, n = 3, \lambda = (2, 1)$. We have $\mathscr{I}_\lambda = \{I_{211} \geqslant I_{121} \geqslant I_{112}\}$ and

$$\begin{pmatrix} \xi_{211} \\ \xi_{121} \\ \xi_{112} \end{pmatrix} = \begin{pmatrix} 1 & 0 & 0 \\ c(z_{21}, P) & \bar{b}(z_{21}) & 0 \\ c(z_{31}, P)\ \bar{b}(z_{31})c(z_{32}, q^2\Pi^*) & \bar{b}(z_{31})\bar{b}(z_{32}) & \end{pmatrix} \begin{pmatrix} v_{211} \\ v_{121} \\ v_{112} \end{pmatrix},$$

where $z_{ij} = z_i/z_j$. On the other hand from (6.2.4) one has

$$\widehat{W}_{\rm id}(z, \Pi^*) = \begin{pmatrix} \widetilde{W}_{I_{211}}(z_{I_{211}}, z, \Pi^*) & 0 & 0 \\ \widetilde{W}_{I_{211}}(z_{I_{121}}, z, \Pi^*) & \widetilde{W}_{I_{121}}(z_{I_{121}}, z, \Pi^*) & 0 \\ \widetilde{W}_{I_{211}}(z_{I_{112}}, z, \Pi^*) & \widetilde{W}_{I_{121}}(z_{I_{112}}, z, \Pi^*) & \widetilde{W}_{I_{112}}(z_{I_{112}}, z, \Pi^*) \end{pmatrix}$$

$$= \begin{pmatrix} 1 & 0 & 0 \\ \frac{\theta(q^{-2}z_{21}\Pi^*)\theta(q^2)}{\theta(q^2 z_{21})\theta(q^{-2}\Pi^*)} & \frac{\theta(z_{21})}{\theta(q^2 z_{21})} & 0 \\ \frac{\theta(q^{-2}z_{31}\Pi^*)\theta(q^2)}{\theta(q^2 z_{31})\theta(q^{-2}\Pi^*)} & \frac{\theta(z_{31})}{\theta(q^2 z_{31})}\frac{\theta(z_{32}\Pi^*)\theta(q^2)}{\theta(q^2 z_{32})\theta(\Pi^*)} & \frac{\theta(z_{31})}{\theta(q^2 z_{31})}\frac{\theta(z_{32})}{\theta(q^2 z_{32})} \end{pmatrix}. \qquad \square$$

7.3 Action of the Elliptic Currents

Now let us consider the action of $\mathscr{U}$ on the GT basis. Note that thanks to Proposition 7.2.2 it suffices to construct an action of $\Delta'^{(n-1)}(L^{\pm}(w))$ on $\xi_{I^{max}}$. From Proposition 7.1.2 and (2.7.7) we obtain the following level-0 action of the dynamical half currents of $\mathscr{U}$ on the GT basis.

Theorem 7.3.1 *Let* $\{\xi_I \mid I \in \mathcal{I}_\lambda, \lambda = (\lambda_1, \lambda_2) \in \mathbb{N}^2, |\lambda| = n\}$ *be the GT basis of* $\widehat{V}_{z_1}\widetilde{\otimes}\cdots\widetilde{\otimes}\widehat{V}_{z_n}$. *Under the abbreviation* $\widetilde{k}_j^{\pm}(w) = (\pi_{z_1}\otimes\cdots\otimes\pi_{z_n})\Delta'^{(n-1)}(\widetilde{k}_j^{\pm}(w))$, $e^{\pm}(w,\Pi^*) = (\pi_{z_1}\otimes\cdots\otimes\pi_{z_n})\Delta'^{(n-1)}(e^{\pm}(w,\Pi^*))$ *and* $f^{\pm}(w,\Pi^*) = (\pi_{z_1}\otimes\cdots\otimes\pi_{z_n})\,\Delta'^{(n-1)}(f^{\pm}(w,\Pi^*))$, *we have*

$$\widetilde{k}_1^{\pm}(w)\xi_I = \prod_{b\in I_2} \left.\frac{\theta(q^{-2}w/z_b)}{\theta(w/z_b)}\right|_{\pm} \xi_I, \quad \widetilde{k}_2^{\pm}(w)\xi_I = \prod_{a\in I_1} \left.\frac{\theta(w/z_b)}{\theta(q^2w/z_b)}\right|_{\pm} \xi_I, \tag{7.3.1}$$

$$e^{\pm}(w,\Pi^*)\xi_I = \sum_{i\in I_2} \left.\frac{\theta(\Pi^* z_i/w)\theta(q^2)}{\theta(\Pi^*)\theta(w/z_i)}\right|_{\pm} \prod_{\substack{k\in I_2\\ \neq i}} \frac{\theta(q^2 z_k/z_i)}{\theta(z_k/z_i)}\, \xi_{I^{i\prime}}, \tag{7.3.2}$$

$$f^{\pm}(w,\Pi^*)\xi_I = \sum_{i\in I_1} \left.\frac{\theta(\Pi^* q^{2(2\lambda_1-n-1)}w/z_i)\theta(q^2)}{\theta(\Pi^* q^{2(2\lambda_1-n-1)})\theta(w/z_i)}\right|_{\pm} \prod_{\substack{k\in I_1\\ \neq i}} \frac{\theta(q^2 z_i/z_k)}{\theta(z_i/z_k)}\, \xi_{I^{\prime i}}, \tag{7.3.3}$$

where $I = (I_1, I_2)$, *and* $I^{i\prime} \in \mathcal{I}_{(\lambda_1+1,\lambda_2-1)}$ *and* $I^{\prime i} \in \mathcal{I}_{(\lambda_1-1,\lambda_2+1)}$ *are defined by*

$$(I^{i\prime})_1 = I_1 \cup \{i\}, \quad (I^{i\prime})_2 = I_2 - \{i\},$$
$$(I^{\prime i})_1 = I_1 - \{i\}, \quad (I^{\prime i})_2 = I_2 \cup \{i\}.$$

The symbols $|_{\pm}$ *specify the expansion directions, for any* w

$$\left.\frac{\theta(wz)(p;p)_\infty^3}{\theta(w)\theta(z)}\right|_{+} = -\sum_{n\in\mathbb{Z}} \frac{1}{1-wp^n} z^n = -\sum_{l\in\mathbb{Z}_{\geq 0}} \left(\frac{w^l}{1-p^l z} - \frac{w^{-l-1}p^{l+1}/z}{1-p^{l+1}/z}\right)$$
$$|p| < |z| < 1,$$

$$\left.\frac{\theta(wz)(p;p)_\infty^3}{\theta(w)\theta(z)}\right|_{-} = \frac{\theta(1/wz)(p;p)_\infty^3}{\theta(1/w)\theta(1/z)}$$
$$= \sum_{n\in\mathbb{Z}} \frac{1}{1-w^{-1}p^n} z^{-n} = \sum_{l\in\mathbb{Z}_{\geq 0}} \left(\frac{w^{-l}p^l/z}{1-p^l/z} - \frac{w^{l+1}}{1-p^{l+1}z}\right)$$
$$1 < |z| < |p^{-1}|.$$

Proof One can check that these satisfy the dynamical version of the relations in Proposition 2.5.2 at level 0. See [100]. □

Example 7.2 Let us consider the case $n = 4$. We here show a calculation using (7.1.1). Noting $\xi_{2211} = v_{2211}$.

$$\begin{aligned}
&k_2^+(w)\xi_{2211} \\
&= (\pi_{z_1}\widetilde{\otimes}\pi_{z_2}\widetilde{\otimes}\pi_{z_3}\widetilde{\otimes}\pi_{z_4})\Delta'^{(3)}(L_{22}^+(w))v_{2211} \\
&= \sum_{k_1,k_2,k_3} \pi_{z_1}(L_{k_3,2}^+(w))v_2\widetilde{\otimes}\pi_{z_2}(L_{k_2,k_3}^+(w))v_2\widetilde{\otimes}\pi_{z_3}(L_{k_1,k_2}^+(w))v_1\widetilde{\otimes}\pi_{z_4}(L_{2,k_1}^+(w))v_1 \\
&= \bar{b}(w/z_3)\bar{b}(w/z_4)\xi_{2211}.
\end{aligned} \tag{7.3.4}$$

Similarly,

$$\begin{aligned}
&(\pi_{z_1}\widetilde{\otimes}\pi_{z_2}\widetilde{\otimes}\pi_{z_3}\widetilde{\otimes}\pi_{z_4})\Delta'^{(3)}(L_{12}^+(w))\xi_{2211} \\
&= \sum_{k_1,k_2,k_3} \pi_{z_1}(L_{k_3,2}^+(w))v_2\widetilde{\otimes}\pi_{z_2}(L_{k_2,k_3}^+(w))v_2\widetilde{\otimes}\pi_{z_3}(L_{k_1,k_2}^+(w))v_1\widetilde{\otimes}\pi_{z_4}(L_{1,k_1}^+(w))v_1 \\
&= v_2\widetilde{\otimes}v_2\widetilde{\otimes}c(w/z_3,\Pi^*)v_2\widetilde{\otimes}v_1 + v_2\widetilde{\otimes}v_2\widetilde{\otimes}\bar{b}(w/z_3)v_1\widetilde{\otimes}c(w/z_4,\Pi^*)v_2 \\
&= c(w/z_3, q^{2(h^{(1)}+h^{(2)})}\Pi^*)v_2\widetilde{\otimes}v_2\widetilde{\otimes}v_2\widetilde{\otimes}v_1 \\
&\qquad +\bar{b}(w/z_3)c(w/z_4, q^{2(h^{(1)}+h^{(2)}+h^{(3)})}\Pi^*)v_2\widetilde{\otimes}v_2\widetilde{\otimes}v_1\widetilde{\otimes}v_2 \\
&= c(w/z_3, q^{-4}\Pi^*)\xi_{2221} + \bar{b}(w/z_3)c(w/z_4, q^{-2}\Pi^*)v_{2212}.
\end{aligned} \tag{7.3.5}$$

The third equality follows from the property (3.3.2). Note that $p^* = p$ on the level-0 representation. In order to convert v_{2212} to ξ_{2212}, remember the definition of the GT basis.

$$\begin{aligned}
\xi_{2212} &= \widetilde{S}_3(\Pi^*)\xi_{2221} \\
&= \mathsf{P}^{(34)}\overline{R}^{(34)}(z_{43}, \Pi^* q^{2(h^{(1)}+h^{(2)})})s_3^z v_{2221} \\
&= c(z_{43}, \Pi^* q^{-4})\xi_{2221} + \bar{b}(z_{43})v_{2212}.
\end{aligned}$$

Therefore one has

$$v_{2212} = \frac{1}{\bar{b}(z_{43})}\xi_{2212} - \frac{c(z_{43}, \Pi^* q^{-4})}{\bar{b}(z_{43})}\xi_{2221}.$$

Substituting this into (7.3.5), one obtains

$$\begin{aligned}
&(\pi_{z_1}\widetilde{\otimes}\pi_{z_2}\widetilde{\otimes}\pi_{z_3}\widetilde{\otimes}\pi_{z_4})\Delta'^{(3)}(L_{12}^+(w))\xi_{2211} \\
&\quad = \frac{\bar{b}(w/z_4)c(w/z_3, q^{-2}\Pi^*)}{\bar{b}(z_{34})}\xi_{2221} + \frac{\bar{b}(w/z_3)c(w/z_4, q^{-2}\Pi^*)}{\bar{b}(z_{43})}\xi_{2212}.
\end{aligned}$$

Here one needs to use the identity

$$c(w/z_3, \Pi^* q^{-4}) - \frac{\bar{b}(w/z_3)c(w/z_4, \Pi^* q^{-2})c(z_{43}, \Pi^* q^{-4})}{\bar{b}(z_{43})} = \frac{\bar{b}(w/z_4)c(w/z_3, \Pi^* q^{-2})}{\bar{b}(z_{34})}.$$

Hence one obtains

$$\begin{aligned} f_2^+(w)\xi_{2211} &= (\pi_{z_1}\widetilde{\otimes}\pi_{z_2}\widetilde{\otimes}\pi_{z_3}\widetilde{\otimes}\pi_{z_4})\Delta'^{(3)}(L_{12}^+(w)L_{22}^+(w)^{-1})\xi_{2211} \\ &= \frac{c(w/z_3, q^{-2}\Pi^*)}{\bar{b}(w/z_3)\bar{b}(z_{34})}\xi_{2221} + \frac{c(w/z_4, q^{-2}\Pi^*)}{\bar{b}(w/z_4)\bar{b}(z_{43})}\xi_{2212}. \end{aligned} \tag{7.3.6}$$

By a similar calculation one obtains

$$\begin{aligned} &(\pi_{z_1}\widetilde{\otimes}\pi_{z_2}\widetilde{\otimes}\pi_{z_3}\widetilde{\otimes}\pi_{z_4})\Delta'^{(3)}(L_{21}^+(w))\xi_{2211} \\ &= \frac{\bar{c}(w/z_2, \Pi^*)}{\bar{b}(z_{12})}\bar{b}(w/z_3)\bar{b}(w/z_4)\xi_{2111} + \frac{\bar{c}(w/z_1, \Pi^*)}{\bar{b}(z_{21})}\bar{b}(w/z_3)\bar{b}(w/z_4)\xi_{1211}, \end{aligned} \tag{7.3.7}$$

$$\begin{aligned} (\pi_{z_1}\widetilde{\otimes}\pi_{z_2}\widetilde{\otimes}\pi_{z_3}\widetilde{\otimes}\pi_{z_4})\Delta'^{(3)}(L_{22}^+(w))\xi_{2111} &= \bar{b}(w/z_2)\bar{b}(w/z_3)\bar{b}(w/z_4)\xi_{2111}, \\ (\pi_{z_1}\widetilde{\otimes}\pi_{z_2}\widetilde{\otimes}\pi_{z_3}\widetilde{\otimes}\pi_{z_4})\Delta'^{(3)}(L_{22}^+(w))\xi_{1211} &= \bar{b}(w/z_1)\bar{b}(w/z_3)\bar{b}(w/z_4)\xi_{1211}. \end{aligned} \tag{7.3.8}$$

The last equation follows from $\xi_{1211} = \widetilde{S}_1(\Pi^*)\xi_{2111}$ and Proposition 7.2.2. Noting also

$$\begin{aligned} \Delta'^{(3)}(L_{22}^+(w)^{-1})\Delta'^{(3)}(g(\Pi^*)) &= \Delta'^{(3)}(L_{22}^+(w)^{-1})(g(\Pi^*)\widetilde{\otimes}1\widetilde{\otimes}1\widetilde{\otimes}1) \\ &= \Delta'^{(3)}(g(q^2\Pi^*))\Delta'^{(3)}(L_{22}^+(w)^{-1}) \end{aligned}$$

for $g(\Pi^*) \in \mathbb{F}$, one obtains

$$\begin{aligned} e_2^+(w)\xi_{2211} &= (\pi_{z_1}\widetilde{\otimes}\pi_{z_2}\widetilde{\otimes}\pi_{z_3}\widetilde{\otimes}\pi_{z_4})\Delta'^{(3)}(L_{22}^+(w)^{-1}L_{21}^+(w))\xi_{2211} \\ &= \frac{\bar{c}(w/z_1, q^2\Pi^*)}{\bar{b}(w/z_1)\bar{b}(z_{21})}\xi_{1211} + \frac{\bar{c}(w/z_2, q^2\Pi^*)}{\bar{b}(w/z_2)\bar{b}(z_{12})}\xi_{2111}. \end{aligned}$$

This is consistent to (7.3.2) by (2.7.9). □

Finally in order to obtain the action of the elliptic currents, note the formula

$$\left.\frac{\theta(wz)}{\theta(w)\theta(z)}\right|_+ - \left.\frac{\theta(wz)}{\theta(w)\theta(z)}\right|_- = \frac{1}{(p;p)_\infty^3}\delta(z) \tag{7.3.9}$$

and Propositions 2.5.1, 2.6.2, and 2.7.1. Then one obtains the following statement.

Corollary 7.3.2 *The level-0 action of the elliptic currents of $\mathscr{U}$ on the GT basis is given by*

$$\psi^{\pm}(w)\xi_I = \varsigma \prod_{a\in I_1} \frac{\theta(q^2 w/z_a)}{\theta(w/z_a)}\Bigg|_{\pm} \prod_{b\in I_2} \frac{\theta(q^{-2} w/z_b)}{\theta(w/z_b)}\Bigg|_{\pm} e^{-Q}\xi_I,$$

$$e(w)\xi_I = -\frac{a^*\theta(q^2)}{(p;p)_\infty^3}\sum_{i\in I_2}\delta(z_i/w)\prod_{\substack{k\in I_2\\ \neq i}}\frac{\theta(q^2 z_k/z_i)}{\theta(z_k/z_i)}\, e^{-Q}\xi_{I^{i\prime}},$$

$$f(w)\xi_I = -\frac{a\theta(q^2)}{(p;p)_\infty^3}\sum_{i\in I_1}\delta(z_i/w)\prod_{\substack{k\in I_1\\ \neq i}}\frac{\theta(q^2 z_i/z_k)}{\theta(z_i/z_k)}\,\xi_{I^{\prime i}}.$$

Proof One can directly check that these satisfy the defining relations of the level-0 $U_{q,p}(\widehat{\mathfrak{sl}}_2)$. See [100]. □

Remark 7.1 In the trigonometric and non-dynamical limit, the combinatorial structure of these formulas is the same as those $N = 2$ case of the geometric representation of $U_q(\widehat{\mathfrak{sl}}_N)$ on the equivariant K-theory of the quiver variety of type A_{N-1} obtained by Ginzburg and Vasserot [64, 159], and by Nakajima [126].

Proposition 7.3.3 *The finite dimensional representation given in Corollary 7.3.2 is the irreducible highest weight representation with the highest weight vector $\xi_{11\cdots 1}$. The elliptic analogue of the Drinfeld polynomials of this representation is given by*

$$P_1(w) = \prod_{a=1}^{n}\theta(q^2 w/z_a). \tag{7.3.10}$$

Proof The statement follows from a similar argument to Theorem 4.11 in [97] and

$$e(w)\xi_{11\cdots 1} = 0 \qquad (j = 1, \cdots, N-1),$$

$$\psi^{+}(w)\xi_{11\cdots 1} = \varsigma\prod_{a=1}^{n}\frac{\theta(q^2 w/z_a)}{\theta(w/z_a)}\xi_{11\cdots 1}.$$ □

Chapter 8
Elliptic q-KZ Equation

We consider a trace of the n-point operator $\phi_{\mu_1\cdots\mu_n}(z_1, \cdots, z_n)$ over the level-1 highest weight representations and show that it satisfies the face type, i.e. dynamical elliptic q-KZ equation. A key to this is a cyclic property of trace and the exchange relation of the vertex operators. Evaluating the trace explicitly we also give an elliptic hypergeometric integral solution to the equation [99].

8.1 Trace of Vertex Operators

Let us consider the n-point operator $\phi_{\mu_1\cdots\mu_n}(z_1, \cdots, z_n)$ in (6.1.2) and the associated partition I of $[1, n]$. We take a trace over the subspace $\mathscr{F}_{a,\nu} = \mathscr{F}_{a,\nu}(l, m)$ of the level-1 $\mathscr{U}$-module $\mathscr{V}(\Lambda_a + \nu, \nu)$ constructed in Sect. 4.3. From (5.1.11), one finds that $\phi_{\mu_1\cdots\mu_n}(z_1, \cdots, z_n)$ preserves $\mathscr{F}_{a,\nu}$ if and only if $n = 2\lambda_1$, i.e. the zero-weight condition $\sum_{k=1}^{n} \bar{\epsilon}_{\mu_k} = 0$ is satisfied.

For $\kappa \in \mathbb{C}^\times$, let us consider the following n-point function.

$$F^a(z_1, \cdots, z_n; \Pi) := \mathrm{tr}_{\mathscr{F}_{a,\nu}}(q^{-\kappa\hat{d}}\, \Phi(z_1) \cdots \Phi(z_n))$$
$$= \sum_{\mu_1, \cdots, \mu_n} v_{\mu_n} \widetilde{\otimes} \cdots \widetilde{\otimes} v_{\mu_1} F^a_{\mu_1\cdots\mu_n}(z_1, \cdots, z_n; \Pi)$$

$$F^a_{\mu_1\cdots\mu_n}(z_1, \cdots, z_n; \Pi) : = \mathrm{tr}_{\mathscr{F}_{a,\nu}}(q^{-\kappa\hat{d}} \Phi^{(a,1-a)}_{\mu_1}(z_1, \Pi) \cdots \Phi^{(1-a,a)}_{\mu_n}(z_n, \Pi)).$$

Here we write the dependence of the vertex operators on the dynamical parameters explicitly $\Phi_\mu(z) = \Phi_\mu(z, \Pi)$.

We show that $F^a_{\mu_1\cdots\mu_n}$ satisfies the elliptic dynamical q-KZ equation. The following lemma is essential.

H. Konno, *Elliptic Quantum Groups*, SpringerBriefs in Mathematical Physics 37,
https://doi.org/10.1007/978-981-15-7387-3_8

Lemma 8.1.1

(1) $F^a_{\mu_1\mu_2\cdots\mu_n}(z_1, z_2, \cdots, q^\kappa z_n; \Pi) = F^{1-a}_{\mu_n\mu_1\cdots\mu_{n-1}}(z_n, z_1, \cdots, z_{n-1}; \Pi q^{-2\langle\bar{\epsilon}_{\mu_n}, h\rangle})$,

(2) $F^a_{\cdots\mu_i\mu_{i+1}\cdots}(\cdots, z_i, z_{i+1}, \cdots; \Pi)$

$$= \sum_{\mu'_i\mu'_{i+1}} R\left(\frac{z_{i+1}}{z_i}, \Pi q^{2\sum_{j=1}^{i-1}\langle\bar{\epsilon}_{\mu_j}, h\rangle}\right)_{\mu_i\mu_{i+1}}^{\mu'_i\mu'_{i+1}} F^a_{\cdots\mu'_{i+1}\mu'_i\cdots}(\cdots, z_{i+1}, z_i, \cdots; \Pi).$$

Proof

(1) Using the cyclic property of trace and

$$\Pi\Phi_{\mu_j}(z, \Pi) = \Phi_{\mu_j}(z, \Pi)\Pi q^{-2\langle\bar{\epsilon}_{\mu_j}, h\rangle}, \tag{8.1.1}$$

one finds

$$LHS = \mathrm{tr}_{\mathscr{F}_{1-a,v}}(\Phi^{(1-a,a)}_{\mu_n}(q^\kappa z_n, \Pi q^{2\sum_{j=1}^{n-1}\langle\bar{\epsilon}_{\mu_j}, h\rangle})q^{-\kappa\hat{d}}$$
$$\times\Phi^{(a,1-a)}_{\mu_1}(z_1, \Pi q^{-2\langle\bar{\epsilon}_{\mu_n}, h\rangle})\cdots\Phi^{(a,1-a)}_{\mu_{n-1}}(z_{n-1}, \Pi q^{-2\langle\bar{\epsilon}_{\mu_n}, h\rangle})).$$

Then using the zero-weight condition $\sum_{j=1}^n \bar{\epsilon}_{\mu_j} = 0$ and $q^{\kappa\hat{d}}\Phi_\mu(z)q^{-\kappa\hat{d}} = \Phi_\mu(q^{-\kappa}z)$, one obtains the desired result.

(2) Use the exchange relations (5.2.4) and (8.1.1). □

By using the properties (1) and (2), we obtain the following statement.

Theorem 8.1.2 *$F^a_{\mu_1\cdots\mu_n}(z_1, \cdots, z_n; \Pi)$ satisfies the dynamical elliptic q-KZ equation*

$$F^a(z_1, \cdots, q^\kappa z_i, \cdots, z_n; \Pi)$$
$$= R'^{(i+1i)}\left(\frac{q^{-\kappa}z_{i+1}}{z_i}, \Pi q^{2\sum_{k=1}^{i-1}h'^{(k)}}\right)\cdots R'^{(ni)}\left(\frac{q^{-\kappa}z_n}{z_i}, \Pi q^{2\sum_{k=1, \neq i}^{n-1}h'^{(k)}}\right)$$
$$\times\Gamma'_i R'^{(1i)}\left(\frac{z_1}{z_i}, \Pi\right)\cdots R'^{(i-1i)}\left(\frac{z_{i-1}}{z_i}, \Pi q^{2\sum_{k=1}^{i-2}h'^{(k)}}\right)F^{1-a}(z_1, \cdots, z_i, \cdots, z_n; \Pi).$$

Here $R'^{(ij)}(z, \Pi) = R^{(n-i+1n-j+1)}(z, \Pi)$ and $h'^{(i)} = h^{(n-i+1)}$ so that $R'^{(ij)}(z, \Pi)$ (resp. $h'^{(i)}$) acts on $\widehat{V}_{z_i}$ and $\widehat{V}_{z_j}$ (resp. on $\widehat{V}_{z_i}$) in $\widehat{V}_{z_n}\widetilde{\otimes}\cdots\widetilde{\otimes}\widehat{V}_{z_1}$. Similarly Γ'_i denotes a shift operator

$$\Gamma'_i f(\,\cdot\,; \Pi) = f(\,\cdot\,; \Pi q^{-2\langle\bar{\epsilon}_\mu, h\rangle})$$

if $q^{2h'^{(i)}}f(\,\cdot\,; \Pi) = q^{2\langle\bar{\epsilon}_\mu, h\rangle}f(\,\cdot\,; \Pi)$ for $f(\,\cdot\,; \Pi) = \sum v_{\mu_n}\widetilde{\otimes}\cdots\widetilde{\otimes}v_{\mu_1}\widetilde{\otimes}f_{\mu_1\cdots\mu_n}(\,\cdot\,; \Pi)$.

8.2 Evaluation of the Trace

Now let us evaluate the trace explicitly. Applying the formula in Theorem 6.2.1, one only needs to take a trace of $\widetilde{\Phi}(t,z)$. Hence one obtains the following result.

Theorem 8.2.1

$$F^a_{\mu_1\cdots\mu_n}(z_1,\cdots,z_n;\Pi)=\oint_{C^{\lambda_1}}\underline{dt}\,\Phi(t,z)\,\omega_{\mu_1\cdots\mu_n}(t,z,\Pi), \tag{8.2.1}$$

$$\begin{aligned}\Phi(t,z)&=\mathrm{tr}_{\mathscr{F}_{a,\nu}}\left(q^{-\kappa d}\,\widetilde{\Phi}(t,z)\right)\\&=C_n\prod_{k=1}^{n}\left((-qz_k)^{-\lambda_1+h/2}\prod_{1\le k\ne l\le n}\frac{\Gamma(q^2z_k/z_l;p,q^\kappa,q^4)}{\Gamma(q^4z_k/z_l;p,q^\kappa,q^4)}\right)\\&\quad\times\prod_{a=1}^{\lambda_1}\left(t_a^{\lambda_1-h}\prod_{b=1}^{n}\frac{\Gamma(t_a/z_b;p,q^\kappa)}{\Gamma(p^*t_a/z_b;p,q^\kappa)}\right)\prod_{1\le a<b\le\lambda_1}\frac{\Gamma(p^*t_a/t_b,p^*t_b/t_a;p,q^\kappa)}{\Gamma(t_a/t_b,t_b/t_a;p,q^\kappa)},\end{aligned}$$

where

$$C_n=(-)^{\frac{1}{2}\lambda_1(\lambda_1-1)}\left(\frac{(p;p)_\infty}{(q^2;p)_\infty}\frac{\Gamma(p;p,q^\kappa)}{\Gamma(q^2;p,q^\kappa)}\right)^{\lambda_1}\left(\frac{(q^4;p,q^4)_\infty}{(q^2;p,q^4)_\infty}\frac{\Gamma(q^2;p,q^\kappa,q^4)}{\Gamma(q^4;p,q^\kappa,q^4)}\right)^n.$$

Proof In $\widetilde{\Phi}(t,z)$, taking normal ordering further between $\Phi_2(z)$'s and $f(t)$'s, we obtain

$$\begin{aligned}\widetilde{\Phi}(t,z)&=\,:\Phi_2(z_1)\cdots\Phi_2(z_n)f(t_1)\cdots f(t_{\lambda_1}):\\&\quad\times\prod_{1\le l<m\le n}<\Phi_2(z_l)\Phi_2(z_m)>^{Sym}\prod_{1\le a<b\le\lambda_1}<f(t_a)f(t_b)>^{Sym}\\&\quad\times\prod_{l=1}^{n}\prod_{a=1}^{\lambda_1}<\Phi_2(z_l)f(t_a)>,\end{aligned}$$

where

$$<\Phi_2(z)f(t)>=(-qz)^{-1}\frac{(pq^{-2}t/z;p)_\infty}{(t/z;p)_\infty}.$$

Then using the formulas in Theorem 5.1.2 and (4.3.6), one can evaluate the trace of the normal ordered operator $:\Phi_2(z_1)\cdots f(t_{\lambda_1}):$ in $\widetilde{\Phi}(t,z)$. See Appendix E. Then combining the result with all the OPE coefficients in $\widetilde{\Phi}(t,z)$ we obtain the desired result. □

Remark 8.1 The integrand of (8.2.1) is a single valued function of $t_1,\cdots,t_{\lambda_1}$.

Remark 8.2 To realize the vertex operator $\Phi(z)$, one can use $F(t_a)$ in (2.3.12) instead of $f(t_a)$ $(a = 1, \cdots, \lambda_1)$ as the elliptic currents [100]. Then in the integrand of $F^a_{\mu_1\cdots\mu_n}(z_1, \cdots, z_n; \Pi)$ one gets an extra factor

$$\prod_{a=1}^{\lambda_1} t_a^{\frac{P+h-1}{r}} = \exp\left\{\frac{\log \Pi \ \log \prod_{a=1}^{\lambda_1} t_a}{\log p}\right\}.$$

A geometric interpretation of such factor is given in 6.1.5 [3]. In addition, in the same paper a role of the vertex function defined in K-theory, which is expressed by a q-hypergeometric integral such as a trigonometric limit of (8.2.1), in 3d SUSY gauge theory is discussed. This and an explicit formula for such function in [105] suggest that the elliptic hypergeometric integral (8.2.1) provides an elliptic analogue of the vertex function with descendent. See Chap. 9 for related geometry.

Remark 8.3 Applications to a calculation of correlation functions in 2d solvable lattice models are discussed in [19, 20, 54, 93, 109, 113].

Chapter 9
Related Geometry

In this chapter, we discuss a geometric interpretation [100] of the results obtained in the previous chapters. Following Aganagic and Okounkov [3], we introduce the equivariant elliptic cohomology $\mathrm{E}_T(X)$ and the elliptic stable envelopes $\mathrm{Stab}_{\mathfrak{C}}$ associated with the cotangent bundle of the Grassmannian variety $X = T^*\mathrm{Gr}(k, n)$. Then we show that the elliptic weight functions in Chap. 6 can be identified with $\mathrm{Stab}_{\mathfrak{C}}$. Based on this identification, we also show a correspondence between the Gelfand-Tsetlin basis (resp. the standard basis) of $\widehat{V}_{z_1}\widetilde{\otimes}\cdots\widetilde{\otimes}\widehat{V}_{z_n}$ in Chap. 7 and the fixed point classes (resp. the stable classes) in $\mathrm{E}_T(X)$. This correspondence allows us to construct an action of $U_{q,p}(\widehat{\mathfrak{sl}}_2)$ on $\mathrm{E}_T(X)$.

9.1 Quiver Varieties

Let us start by recalling some basic properties of the quiver varieties [124, 125]. We refer to [85] for details.

Let Q be a finite quiver with vertex set I and edges E. For $\mathbf{v} = (v_1, \cdots, v_{|\mathsf{I}|})$, $\mathbf{w} = (w_1, \cdots, w_{|\mathsf{I}|}) \in \mathbb{N}^{|\mathsf{I}|}$ consider the vector spaces V_i, W_i with $\dim V_i = v_i$, $\dim W_i = w_i$ $(i = 1, \cdots, |\mathsf{I}|)$. We define

$$R(\mathbf{v}, \mathbf{w}) = \bigoplus_{(i\to j)\in\mathsf{E}} \mathrm{Hom}(V_i, V_j) \oplus \bigoplus_{i\in\mathsf{I}} \mathrm{Hom}(V_i, W_i) \qquad (9.1.1)$$

and consider

$$T^*R(\mathbf{v}, \mathbf{w}) = R(\mathbf{v}, \mathbf{w}) \oplus R(\mathbf{v}, \mathbf{w})^{\vee}.$$

H. Konno, *Elliptic Quantum Groups*, SpringerBriefs in Mathematical Physics 37,
https://doi.org/10.1007/978-981-15-7387-3_9

Here

$$R(\mathbf{v},\mathbf{w})^\vee = \bigoplus_{(i\to j)\in \mathrm{E}} \mathrm{Hom}(V_j, V_i) \oplus \bigoplus_{i\in \mathrm{I}} \mathrm{Hom}(W_i, V_i).$$

Denote a point in $T^*R(\mathbf{v},\mathbf{w})$ by $(\mathbf{x},\mathbf{y},\mathbf{i},\mathbf{j})$, where $\mathbf{x} = \oplus_{(i\to j)}x_{ij}$, $\mathbf{y} = \oplus_{(i\to j)}y_{ij}$, $\mathbf{i} = \oplus_i \mathbf{i}_i$, $\mathbf{j} = \oplus_i \mathbf{j}_i$,

$$x_{ij} \in \mathrm{Hom}(V_i, V_j), \quad y_{ij} \in \mathrm{Hom}(V_j, V_i), \quad \mathbf{i}_i \in \mathrm{Hom}(W_i, V_i), \quad \mathbf{j}_i \in \mathrm{Hom}(V_i, W_i).$$

On $T^*R(\mathbf{v},\mathbf{w})$, $G(\mathbf{v}) = \prod_{i=1}^{|\mathrm{I}|} GL(v_i)$ acts by

$$g\cdot(\mathbf{x},\mathbf{y},\mathbf{i},\mathbf{j}) = (g\mathbf{x}g^{-1}, g\mathbf{y}g^{-1}, g\mathbf{i}, \mathbf{j}g^{-1}).$$

It is Hamiltonian and the moment map $\mu : T^*R(\mathbf{v},\mathbf{w}) \to \mathfrak{g}(\mathbf{v})^\vee$ is given by

$$\mu(\mathbf{x},\mathbf{y},\mathbf{i},\mathbf{j}) = [\mathbf{x},\mathbf{y}] + \mathbf{ij}.$$

Here $\mathfrak{g}(\mathbf{v}) = \mathrm{Lie}\, G(\mathbf{v})$ and we identify $\mathfrak{g}(\mathbf{v}) \cong \mathfrak{g}(\mathbf{v})^\vee$ via the Killing form. Take a stability parameter $\theta \in \mathbb{Z}^{|\mathrm{I}|}$ corresponding to a character $\chi : G(\mathbf{v}) \mapsto \prod_i (\det g_i)^{-\theta_i}$.

Definition 9.1.1 ([124]) The quiver variety $\mathcal{M}_\theta(\mathbf{v},\mathbf{w})$ is defined to be the twisted GIT quotient

$$\mathcal{M}_\theta(\mathbf{v},\mathbf{w}) = \mu^{-1}(0) /\!\!/_\theta\, G(\mathbf{v}).$$

A quiver variety $\mathcal{M}_\theta(\mathbf{v},\mathbf{w})$ admits a torus action $T = A \times \mathbb{C}_\hbar^\times$, $A = \prod_i (\mathbb{C}^\times)^{w_i}$. The part A acts on the framing space $\oplus_i W_i$, whereas $\mathbb{C}_\hbar^\times$ acts on the cotangent fiber by

$$\hbar\cdot(\mathbf{x},\mathbf{y},\mathbf{i},\mathbf{j}) = (\mathbf{x}, \hbar^{-1}\mathbf{y}, \hbar^{-1}\mathbf{i}, \mathbf{j}).$$

Let $S \subset G(\mathbf{v})$ be the maximal torus and consider the projection $\mu_S := \pi_S \circ \mu$ by $\pi_S : \mathfrak{g}(\mathbf{v})^\vee \to (\mathrm{Lie}\, S)^\vee$.

Definition 9.1.2 The abelian quotient

$$\mu_S^{-1}(0) /\!\!/_\theta\, S$$

becomes a hypertoric variety and is called the abelianization of $\mathcal{M}_\theta(\mathbf{v},\mathbf{w})$.

9.1.1 Cotangent Bundle of the Grassmannian

Let us consider the Grassmannian variety $\mathrm{Gr}(k, n)$ consisting of all k-dimensional subspaces of $\mathbb{C}^n$. The cotangent bundle $X = T^*\mathrm{Gr}(k, n)$ has the following description as the quiver variety associated with the type A_1 Dynkin quiver with dimension $\mathbf{v} = k$ and framing $\mathbf{w} = n$. Let $R = \mathrm{Hom}(\mathbb{C}^k, \mathbb{C}^n)$ be a vector space of complex $n \times k$ matrices. Then $T^*R = R \oplus R^\vee \cong \mathrm{Hom}(\mathbb{C}^k, \mathbb{C}^n) \oplus \mathrm{Hom}(\mathbb{C}^n, \mathbb{C}^k)$. Let $\mathbf{i} \in R^\vee, \mathbf{j} \in R$. There is an action of $GL(k)$ by

$$g \cdot (\mathbf{i}, \mathbf{j}) = (g\mathbf{i}, \mathbf{j}g^{-1}) \qquad g \in GL(k).$$

We have the Hamiltonian moment map $\mu : T^*R \to \mathfrak{gl}(k)^\vee = \mathrm{Hom}(\mathbb{C}^k, \mathbb{C}^k)$ given by

$$\mu(\mathbf{i}, \mathbf{j}) = \mathbf{ij}.$$

The quiver variety $\mathcal{M}_\theta(\mathbf{v}, \mathbf{w})$ associated with the A_1 quiver is then given by

$$\mathcal{M}_\theta(\mathbf{v}, \mathbf{w}) = \mu^{-1}(0) /\!\!/_\theta GL(k) = \mu^{-1}(0) \cap \{\theta - \text{semistable points}\}/GL(k).$$

Taking $\theta = 1$, one has

$$\mathcal{M}_\theta(\mathbf{v}, \mathbf{w}) = \{(\mathbf{j}, \mathbf{i}) \in T^*R \mid \mathbf{ij} = 0, \mathrm{Ker}\, \mathbf{j} = 0\,\}/GL(k) \cong T^*\mathrm{Gr}(k, n).$$

The special case $k = 1$ of $T^*\mathrm{Gr}(k, n)$ gives $T^*\mathbb{P}(\mathbb{C}^n)$. The abelianization X_S of X by $S = (\mathbb{C}^\times)^k \subset GL(k)$ is then given by

$$X_S = (T^*\mathbb{P}(\mathbb{C}^n))^k.$$

9.2 Equivariant Elliptic Cohomology $\mathrm{Ell}_T(X)$

Let $E = \mathbb{C}^\times/p^{\mathbb{Z}}$ ($|p| < 1$) be an elliptic curve. Let X be a quiver variety endowed with an action of torus $T = (\mathbb{C}^\times)^m$, where $m = \sum_{i=1}^{|I|} w_i + 1$ (Sect. 9.1). We follow [3, 60, 65, 70, 141] for the definition of the T-equivariant elliptic cohomology $\mathrm{Ell}_T(X)$. A basic idea is to formulate $\mathrm{Ell}_T(X)$ by gluing Spec $\mathrm{H}_T^*(X^T)$, which is an affine schemes over Spec $\mathrm{H}_T^*(\mathrm{pt}) \cong \mathrm{Lie}\, T \cong \mathbb{C}^m$. Some basic properties related to $\mathrm{Ell}_T(X)$ are summarized as follows.

(1) (Equivariant elliptic cohomology) The T-equivariant elliptic cohomology Ell_T is a functor from finite T-spaces X to super schemes, covariant in both T and X, satisfying a set of axioms ([70], 4.1 in [60] and 2.2.2 in [3]). The covariance

in T implies that $\mathrm{Ell}_T(X)$ is a scheme over

$$\mathscr{E}_T := \mathrm{Ell}_T(\mathrm{pt}) = T/p^{\mathrm{cochar}(T)} \cong E^m.$$

Here

$$\mathrm{char}(T) = \mathrm{Hom}(T, \mathbb{C}^\times) \cong \mathbb{Z}^m \cong \mathrm{cochar}(T) = \mathrm{Hom}(\mathbb{C}^\times, T).$$

More precisely, due to the functoriality of Ell_T, the projection $X \to \mathrm{pt}$ yields

$$\pi : \mathrm{Ell}_T(X) \to \mathscr{E}_T.$$

Then for a small analytic neighborhood U of a point $t \in T$, one has the diagram

$$\begin{array}{ccccc} \mathrm{Spec}\, \mathrm{H}_T^*(X^{T_t}) & \longleftarrow & \pi^{-1}(U) & \longrightarrow & \mathrm{Ell}_T(X) \\ \downarrow & & \downarrow & & \downarrow \\ \mathbb{C}^m & \longleftarrow & U & \longrightarrow & \mathscr{E}_T. \end{array}$$

(2) (Embedding structure) Due to a construction of X as a quotient by $G(\mathbf{v}) = \prod_{l=1}^{|\mathsf{I}|} GL(v_l)$, there is a collection of tautological vector bundles $\{\mathscr{V}_l\}$ of $\mathrm{rk} = v_l$ $(l = 1, \cdots, |\mathsf{I}|)$ over X associated with the vector representations $\mathbb{C}^{v_l}$ of $\mathrm{GL}(v_l)$. One also has a map

$$\mathrm{Ell}_T(X) \to \mathscr{E}_T \times E^{(v_1)} \times \cdots \times E^{(v_{|\mathsf{I}|})}, \tag{9.2.1}$$

which is an embedding near the origin of $\mathscr{E}_T$ (2.5.2 in [3] and [118]). Here $E^{(m)} = E^m/\mathfrak{S}_m$ denotes the symmetric product of E.

(3) (Polarization bundle) The polarization bundle $T^{1/2}X$ of X is an element of $\mathrm{K}_T(X)$ such that

$$TX = T^{1/2}X + \hbar^{-1} \otimes (T^{1/2}X)^\vee.$$

For $X = \mathscr{M}_\theta(\mathbf{v}, \mathbf{w})$ given in Sect. 9.1, from (9.1.1) one can realize $T^{1/2}X$ as the following virtual vector bundle.

$$T^{1/2}X = \sum_{(i\to j)\in\mathsf{E}} \mathscr{V}_j \otimes \mathscr{V}_i^\vee + \sum_{i\in\mathsf{I}} \mathscr{W}_i \otimes \mathscr{V}_i^\vee - \sum_{i\in\mathsf{I}} \mathscr{V}_i \otimes \mathscr{V}_i^\vee,$$

where $\mathscr{W}_i$ $(i \in \mathsf{I})$ denote topologically trivial rank-w_i vector bundles on X associated with the framing space W_i and carrying the vector representation of $GL(w_i)$. The last term is a contribution from the reduction $/G(\mathbf{v})$. (4.2.2 in [3]). Let $s_a^{(i)}$ $(a = 1, \cdots, v_i)$ denote the Chern roots of $\mathscr{V}_i$ (2.4 in [149] and 3.4 in [134]). One has

$$T^{1/2}X = \sum_{(i\to j)\in \mathsf{E}} \sum_{a=1}^{v_i} \sum_{b=1}^{v_j} \frac{s_b^{(j)}}{s_a^{(i)}} + \sum_{i\in \mathsf{I}} \sum_{a=1}^{v_i} \sum_{b=1}^{w_i} \frac{\mathfrak{a}_b^{(i)}}{s_a^{(i)}} - \sum_{i\in \mathsf{I}} \sum_{\substack{a,b=1 \\ a\neq b}}^{v_i} \frac{s_b^{(i)}}{s_a^{(i)}}, \tag{9.2.2}$$

where $\mathfrak{a}_b^{(i)}$ $(b = 1, \cdots, w_i)$ are equivariant parameters associated with $\mathscr{W}_i$.

(4) (Thom sheaf) The Thom class map $\Theta : \mathrm{K}_T(X) \to \mathrm{Pic}(\mathrm{Ell}_T(X))$ is a map of a T-equivariant complex vector bundle ξ to a line bundle $\Theta(\xi)$ over $\mathrm{Ell}_T(X)$. The line bundle $\Theta(\xi)$ is called the Thom sheaf of ξ. See 2.6.1 in [3] and Definition 6.1 in [60].

(5) (Pushforward) Let $f : X \to Y$ be a holomorphic map of T-spaces. Pullback in the elliptic cohomology is the contravariant functoriality map $\mathrm{Ell}(f) : \mathrm{Ell}_T(X) \to \mathrm{Ell}_T(Y)$ ((1.7.4) in [65] and 2.6.1 in [3]). If f is proper, pushforward is a morphism

$$f_* : \mathrm{Ell}(f)_*\Theta(-N_f) \to \mathscr{O}_{\mathrm{Ell}_T(Y)} \tag{9.2.3}$$

of sheaves on $\mathrm{Ell}_T(Y)$, where $N_f := f^*TY - TX \in \mathrm{K}_T(X)$ is the normal bundle to f. See (2.3.2) in [65] and (21) in [3].

(6) (Picard group) The tautological line bundles $\wedge^{v_i}\mathscr{V}_i$ $(i = 1, \cdots, |\mathsf{I}|)$ generate the Picard group $\mathrm{Pic}(X) \cong \mathbb{Z}^{|\mathsf{I}|}$. We also introduce the T-equivariant Picard group $\mathrm{Pic}_T(X)$ as the extension of $\mathrm{Pic}(X)$

$$0 \to \mathrm{char}(T) \to \mathrm{Pic}_T(X) \to \mathrm{Pic}(X) \to 0. \tag{9.2.4}$$

Let us set

$$\mathscr{E}_{\mathrm{Pic}(X)} := \mathrm{Pic}(X) \otimes_{\mathbb{Z}} E \cong E^{|\mathsf{I}|},$$
$$\mathscr{E}_{\mathrm{Pic}_T(X)} := \mathrm{Pic}_T(X) \otimes_{\mathbb{Z}} E.$$

Due to (9.2.4), one has

$$0 \to \mathscr{E}_T^\vee \to \mathscr{E}_{\mathrm{Pic}_T(X)} \to \mathscr{E}_{\mathrm{Pic}(X)} \to 0. \tag{9.2.5}$$

(7) (Equivariant parameters and Kähler parameters) Let us consider the extension of $\mathrm{Ell}_T(X)$

$$\mathrm{E}_T(X) := \mathrm{Ell}_T(X) \times \mathscr{E}_{\mathrm{Pic}_T(X)}$$

as a scheme over $\mathcal{B}_{T,X} = \mathcal{E}_T \times \mathcal{E}_{\mathrm{Pic}_T(X)}$. The variables in the two factors of $\mathcal{B}_{T,X}$, $\mathfrak{a}_1^{(i)}, \cdots, \mathfrak{a}_{w_i}^{(i)}$ $(1 \leq i \leq |\mathrm{I}|)$, $\hbar$ in $\mathcal{E}_T \cong \prod_i E^{w_i} \times E$ and ζ_j $(1 \leq j \leq |\mathrm{I}|)$ in $\mathcal{E}_{\mathrm{Pic}(X)} \subset \mathcal{E}_{\mathrm{Pic}_T(X)}$, are called the equivariant and the Kähler parameters, respectively. See 2.7.3 in [3]. In Sect. 9.4, we identify ζ_j with the dynamical parameters. See Sect. 9.4. As seen from (9.2.5), there are remaining degree of freedom $\mathcal{E}_T^\vee$ in $\mathcal{E}_{\mathrm{Pic}_T(X)}$. This part gives only a trivial contribution to the stable envelopes discussed below. So we do not touch it further. (3.3.7 in [3])

(8) (Universal line bundles) $\mathcal{E}_{\mathrm{Pic}_T(X)}$ and $\mathcal{E}^\vee_{\mathrm{Pic}_T(X)} := \mathrm{Hom}(\mathrm{Pic}_T(X), E)$ are dual abelian varieties with each other. Hence there exists a universal line bundle $\mathcal{U}_{\text{Poincaré}}$ over $\mathcal{E}^\vee_{\mathrm{Pic}_T(X)} \times \mathcal{E}_{\mathrm{Pic}_T(X)}$. Note that the Chern class ((1.8) in [65]) defined by an equivariant line bundle over X yields a group homomorphism

$$c : \mathrm{Pic}_T(X) \to \mathrm{Maps}(\mathrm{Ell}_T(X) \to E),$$

which can be regarded as a map

$$\tilde{c} : \mathrm{Ell}_T(X) \to \mathcal{E}^\vee_{\mathrm{Pic}_T(X)}$$

(2.7.1 in [3]). One hence obtains a line bundle $\mathcal{U}$ on $\mathrm{E}_T(X)$ by

$$\mathcal{U} = (\tilde{c} \times 1)^* \, \mathcal{U}_{\text{Poincaré}}.$$

It is well-known that the universal line bundle of $E \times E^\vee$ has meromorphic section of the form

$$\frac{\theta(xy)}{\theta(x)\theta(y)}.$$

9.3 Elliptic Stable Envelopes

Following [3], we introduce the elliptic stable envelopes.

9.3.1 Chamber Structure

Let $\mathrm{Hom}_{\mathrm{grp}}(\mathbb{C}^\times, A)$ be a one-parameter subgroup and consider its real form $\mathfrak{a}_\mathbb{R} = \mathrm{Hom}_{\mathrm{grp}}(\mathbb{C}^\times, A) \otimes_\mathbb{Z} \mathbb{R} \subset \mathrm{Lie}\, A$. The space $\mathfrak{a}_\mathbb{R}$ can be decomposed into finitely many chambers $\mathfrak{C}$ defined as a connected component of the compliment of the union of hyperplanes given by ρ such that $X^{\rho(\mathbb{C}^\times)} \neq X^A$ [127].

Let X^A be the A-fixed point locus in X and $X^A = \bigsqcup_I F_I$ a decomposition to connected components. Let $\rho \in \mathfrak{C}$. For every $\mathscr{S} \subset X^A$ we define its attracting set

$$\mathrm{Attr}(\mathscr{S}) = \{(x, s), s \in \mathscr{S}, \lim_{t\to 0} \rho(t)x = s\} \subset X \times X^A,$$

and denote by $\mathrm{Attr}^f(\mathscr{S})$ the full attracting set, which is the minimal closed subset of X that contains the diagonal $\mathscr{S} \times \mathscr{S}$ and is closed under taking $\mathrm{Attr}(\cdot)$. We then define a partial ordering on $\{F_I\}$ by

$$F_J \le F_I \quad \Leftrightarrow \quad \mathrm{Attr}^f(F_I) \cap F_J \ne \emptyset.$$

9.3.2 *Definition*

For a pair (μ, ν), $\mu \in \mathrm{char}(T) = \mathrm{Hom}(\mathscr{E}_T, E)$, $\nu \in \mathrm{Pic}_T(X) = \mathrm{Hom}(E, \mathscr{E}_{\mathrm{Pic}_T(X)})$, let τ denote the automorphism of $\mathscr{B}_{T,X} = \mathscr{E}_T \times \mathscr{E}_{\mathrm{Pic}_T(X)}$

$$\tau(\mu\nu) : (\mathfrak{a}, \hbar, \zeta) \mapsto (\mathfrak{a}, \hbar, \zeta + \nu(\mu(\mathfrak{a}, \hbar))).$$

For each chamber $\mathfrak{C}$ of Lie A, one can decompose $T^{1/2}X|_{X^A}$, the polarization of X restricted to X^A, into

$$T^{1/2}X|_{X^A} = T^{1/2}X|_{X^A,>0} \oplus T^{1/2}X|_{X^A,\text{fixed}} \oplus T^{1/2}X|_{X^A,<0}, \tag{9.3.1}$$

where > 0, fixed, < 0 denotes the attracting, fixed and repelling direction parts. The fixed part defines the polarization of X^A, i.e. $T^{1/2}X^A = T^{1/2}X|_{X^A,\text{fixed}}$. Let us set

$$\mathrm{ind} := T^{1/2}X|_{X^A,>0} \in \mathrm{K}_T(X^A).$$

We have

$$\det \mathrm{ind} \in \mathrm{Pic}_T(X^A)$$

and a translation, for $\hbar \in \mathrm{char}(T)$,

$$\tau(-\hbar \det \mathrm{ind}) : \mathscr{B}_{T,X^A} \to \mathscr{B}_{T,X^A}.$$

Let $\iota : X^A \to X$ be the inclusion map, which is proper. For the universal line bundle $\widetilde{\mathscr{U}}$ on $\mathrm{Ell}_T(X^A) \times \mathscr{E}_{\mathrm{Pic}_T(X)}$ obtained by pulling back by $\mathrm{Ell}(\iota)^* \times 1$ from the one on $\mathrm{E}_T(X)$, we consider the following shifted universal line bundle on $\mathrm{E}_T(X^A)$.

$$\mathscr{U}' = (1 \times \iota)^* \tau(-\hbar \det \mathrm{ind})^* \widetilde{\mathscr{U}}.$$

$$\begin{array}{ccccc} \mathrm{E}_T(X^A) & \xleftarrow{1\times\iota^*} & \mathrm{Ell}_T(X^A)\times\mathscr{E}_{\mathrm{Pic}_T(X)} & \xrightarrow{\mathrm{Ell}(\iota)\times 1} & \mathrm{E}_T(X) \\ \pi\times 1 \downarrow & & \downarrow & & \downarrow \pi\times 1 \\ \mathscr{B}_{T,X^A} & \xleftarrow{1\times\iota^*} & \mathscr{B}_{T,X} & \xrightarrow{1} & \mathscr{B}_{T,X}. \end{array}$$

Definition 9.3.1 The elliptic stable envelop $\mathrm{Stab}_{\mathfrak{C}}$ is defined to be a map of $\mathscr{O}_{\mathscr{B}_{T,X}}$-modules

$$\Theta(T^{1/2}X^A)\otimes\mathscr{U}' \quad \xrightarrow{\mathrm{Stab}_{\mathfrak{C}}} \quad \Theta(T^{1/2}X)\otimes\mathscr{U}\otimes\cdots, \tag{9.3.2}$$

where $\Theta(T^{1/2}X)$ denotes the Thom sheaf of a polarization, and $\cdots$ stands for a certain line bundle pulled back from

$$\mathscr{B}' = \mathscr{B}_{T,X}/\mathscr{E}_A,$$

where $\mathscr{E}_A = \mathrm{Ell}_A(\mathrm{pt})$, i.e. line bundles whose sections are described in terms of $\hbar$ and ζ_i $(i = 1, \cdots, |\mathrm{I}|)$. $\mathrm{Stab}_{\mathfrak{C}}$ is subjected to the following two conditions (3.3.5 in [3]).

(i) (triangularity) Let s_K be an elliptic cohomology class supported on F_K locally over $\mathscr{B}_{T,X}$. Then $\mathrm{Stab}_{\mathfrak{C}}(s_K)$ is supported on $\mathrm{Attr}^f(F_K)$. In particular unless $F_K \geq F_I$ we have

$$\mathrm{Stab}_{\mathfrak{C}}(s_K)|_{F_I} = 0.$$

(ii) (normalization) Near the diagonal in $X\times F_K$, one has

$$\mathrm{Stab}_{\mathfrak{C}} = (-1)^{\mathrm{rk\ ind}}\ j_*\pi^*, \tag{9.3.3}$$

where

$$F_K \xleftarrow{\pi} \mathrm{Attr}(F_K) \xrightarrow{j} X$$

are the natural projection and inclusion maps.

Remark 9.1 In (9.3.2) the reason why one needs the Thom sheaves of the polarizations and the shift in the universal line bundle $\mathscr{U}'$ is due to the requirement (ii). Noting $N_j = j^*TX - T\mathrm{Attr}(F_K) \cong N_{X^A,<0}$, the repelling part of the normal bundle to X^A, one finds that the definition of the pushforward j_* from (9.2.3) leads to

$$\Theta(-N_{X^A,<0}) \ \xrightarrow{j_*\pi^*}\ \mathscr{O}_{\mathrm{Ell}_T(X)}.$$

In order to absorb this contribution into the source and the target, the first attempt could be to tensor the Thom sheaves $\Theta(T^{1/2}X^A)$, $\Theta(T^{1/2}X)$ to them, respectively.

However, this choice turns out not enough to make $\mathrm{Stab}_{\mathfrak{C}}$ free from factors of automorphy with respect to the equivariant parameters $\mathfrak{a}_b^{(i)}$ $(b = 1, \cdots, w_i, i = 1, \cdots, |l|)$ in $\mathscr{E}_A$. The role of the shift $\tau(-\hbar \det \mathrm{ind})$ in $\mathscr{U}'$ is to fill this gap. In fact, one finds (3.3.6 in [3])

$$\Theta(T^{1/2}X - T^{1/2}X^A - N_{X^A,<0})|_{\mathscr{E}_A\text{-orbits}} \cong \frac{\tau(-\hbar \det \mathrm{ind})^* \mathscr{U}}{\mathscr{U}}.$$

$\because$ Noting

$$N_{X^A,<0} \sim TX|_{X^A,<0} \sim T^{1/2}X|_{X^A,<0} + \hbar^{-1}(T^{1/2}X|_{X^A,>0})^\vee,$$

one has

$$T^{1/2}X - T^{1/2}X^A - N_{X^A,<0} \sim T^{1/2}X|_{X^A,>0} - \hbar^{-1}(T^{1/2}X|_{X^A,>0})^\vee.$$

Hence assuming the expression $\mathrm{ind} = T^{1/2}X|_{X^A,>0} = \sum_{i=1}^l w_i$, $l = \mathrm{rk}\ \mathrm{ind}$, in terms of the Chern roots w_i one gets

$$T^{1/2}X - T^{1/2}X^A - N_{X^A,<0} \sim \sum_{i=1}^l w_i - \sum_{i=1}^l \frac{1}{\hbar w_i}.$$

Hence

$$\Theta(T^{1/2}X - T^{1/2}X^A - N_{X^A,<0}) \sim \prod_{i=1}^l \frac{\theta(w_i)}{\theta(\hbar w_i)}.$$

Comparing the quasi-periodicity, one finds that this gives a meromorphic section of

$$\frac{\tau(-\hbar \det \mathrm{ind})^* \mathscr{U}}{\mathscr{U}}. \qquad \square$$

9.3.3 Example: The Case $X = T^*\mathbb{P}(\mathbb{C}^n)$

We follow a construction of $\mathrm{Stab}_{\mathfrak{C}}$ for X given in 3.4 from [3]. Let s be a coordinate on $GL(1)/p^{\mathbb{Z}} = E = E^\vee$. The variable s gives the Chern root of the tautological line bundle $\mathscr{V} = \mathscr{O}(1)$ over X. We denote by ζ the Kähler parameter dual to s.

Let us take a basis $\{e_i\ (i = 1, \cdots, n)\}$ of $\mathbb{C}^n$, on which the torus $A = (\mathbb{C}^\times)^n$ acts as $\mathrm{diag}(\mathfrak{a}_1, \cdots, \mathfrak{a}_n)$. Hence $F_m = \mathbb{C}e_m$ $(m = 1, \cdots, n)$ give the A-fixed points. We chose the chamber $\mathfrak{C}$ such that

$$F_1 > F_2 > \cdots > F_n.$$

The polarization is given by

$$T^{1/2}X = \sum_{i=1}^{n} \mathfrak{a}_i \mathscr{V} - \mathscr{V} \otimes \mathscr{V}^\vee.$$

Hence in terms of the Chern root s the polarization of X is given by

$$\Theta(T^{1/2}X) = \prod_{i=1}^{n} \theta(\mathfrak{a}_i s) \tag{9.3.4}$$

up to a constant factor. When one restricts to F_m, $\mathscr{V}$ becomes the trivial bundle with weight $\mathfrak{a}_m^{-1}$. Hence one has

$$T^{1/2}X|_{F_m} = \sum_{\substack{i=1\\ i\neq m}}^{n} \frac{\mathfrak{a}_i}{\mathfrak{a}_m} = \sum_{i>m} \frac{\mathfrak{a}_i}{\mathfrak{a}_m} + \sum_{i<m} \frac{\mathfrak{a}_i}{\mathfrak{a}_m} \tag{9.3.5}$$

and

$$\operatorname{ind}(F_m) = T^{1/2}X|_{F_m,>0} = \sum_{i>m} \frac{\mathfrak{a}_i}{\mathfrak{a}_m},$$

$$\det \operatorname{ind}(F_m) = \prod_{i>m} \frac{\mathfrak{a}_i}{\mathfrak{a}_m} = \mathfrak{a}_m^{m-n} \prod_{i>m} \mathfrak{a}_i.$$

From *Remark* 9.1, one gets near F_k

$$\frac{\tau(-\hbar \det \operatorname{ind})^* \mathscr{U}}{\mathscr{U}} \cong \frac{\theta(s\mathfrak{a}_m)}{\theta(s\mathfrak{a}_m \hbar^{m-n})} \prod_{i>m} \frac{\theta(s\mathfrak{a}_i)}{\theta(s\mathfrak{a}_i \hbar)} \times \prod \theta(\hbar^{m-n}). \tag{9.3.6}$$

In addition, the Thom class of the repelling part of the normal bundle to F_k in X is given by

$$\begin{aligned} \Theta(N_{F_m,<0}) &= \Theta(T^{1/2}X_{F_m,<0} + \hbar^{-1}(T^{1/2}X_{F_m,>0})^\vee) \\ &= (-)^{n-m} \prod_{i<m} \theta(\mathfrak{a}_i/\mathfrak{a}_m) \prod_{i>m} \theta(\mathfrak{a}_i \hbar/\mathfrak{a}_m). \end{aligned} \tag{9.3.7}$$

Then the elliptic stable envelopes $\mathrm{Stab}_{\mathfrak{C}}(F_m)$ for $X = T^*\mathbb{P}(\mathbb{C}^n)$ is given by

$$\mathrm{Stab}_{\mathfrak{C}}(F_m) = \prod_{1\leq i<m} \theta(s\mathfrak{a}_i) \frac{\theta(s\mathfrak{a}_m \zeta \hbar^{m-n})}{\theta(\zeta \hbar^{m-n})} \prod_{m<i\leq n} \theta(t\mathfrak{a}_i \hbar). \tag{9.3.8}$$

In fact this satisfies the triangularity condition

$$\mathrm{Stab}_{\mathfrak{C}}(F_m)|_{F_i} = 0 \qquad \text{for } i < m \quad \text{i.e.} \quad F_i > F_m, \,.$$

and the normalization

$$\mathrm{Stab}_{\mathfrak{C}}(F_m)|_{F_m} = \prod_{1 \le i < m} \theta(\mathfrak{a}_i/\mathfrak{a}_m) \prod_{m < i \le n} \theta(\mathfrak{a}_i \hbar/\mathfrak{a}_m) = (-)^{n-m} \Theta(N_{F_m,<0}).$$

Note also that one can rewrite (9.3.8) as

$$\mathrm{Stab}_{\mathfrak{C}}(F_m) = \frac{\prod_{i=1}^n \theta(s\mathfrak{a}_i) \times \mathcal{U}}{\frac{\theta(s\mathfrak{a}_m)}{\theta(s\mathfrak{a}_m \hbar^{m-n})} \prod_{i>m} \frac{\theta(s\mathfrak{a}_i)}{\theta(s\mathfrak{a}_i \hbar)} \times \theta(\hbar^{m-n}) \times \mathcal{U}|_{F_m}},$$

where

$$\mathcal{U} = \frac{\theta(s\mathfrak{a}_m \hbar^{m-n} \zeta)}{\theta(s\mathfrak{a}_m \hbar^{m-n})\theta(\zeta)}, \qquad \mathcal{U}|_{F_m} = \frac{\theta(\hbar^{m-n}\zeta)}{\theta(\hbar^{m-n})\theta(\zeta)}.$$

Each factor has the following meaning

- $\prod_{i=1}^n \theta(s\mathfrak{a}_i)$: the polarization in the target from (9.3.4)
- $\mathcal{U}$: universal line bundle in the target
- $\mathcal{U}|_{F_m}$: universal line bundle in the source
- the factors

$$\frac{\theta(s\mathfrak{a}_m)}{\theta(s\mathfrak{a}_m \hbar^{m-n})} \prod_{i>m} \frac{\theta(s\mathfrak{a}_i)}{\theta(s\mathfrak{a}_i \hbar)} \times \theta(\hbar^{m-n})$$

  are from (9.3.6).

9.3.4 Example: The Case $X = T^*\mathrm{Gr}(k, n)$

Let $A = \mathrm{diag}(\mathfrak{a}_1, \cdots, \mathfrak{a}_n)$ and chose the chamber $\mathfrak{C}$ as above. Namely,

$$\mathfrak{a}_j/\mathfrak{a}_i > 0 \Leftrightarrow i < j.$$

The A-fixed point components are labeled by partitions $I \in \mathscr{I}_{(k,n-k)}$ of $[1, n]$. More explicitly let $I = (I_1, I_2)$, $I_1 = \{i_1 < i_2 < \cdots < i_k\} \subset [i, n]$. Then

$$F_I = \mathrm{Span}_{\mathbb{C}}\{e_{i_1}, \cdots, e_{i_k}\}.$$

In order to obtain the stable envelopes for $X = T^*\mathrm{Gr}(k,n)$, we use the abelianization formula developed in [3, 146, 148].

Let $S = (\mathbb{C}^\times)^k$ be the maximal torus of $GL(k)$. The abelianization X_S is given by

$$X_S = \left(T^*\mathbb{P}(\mathbb{C}^n)\right)^k. \tag{9.3.9}$$

We identify coordinates in $\begin{pmatrix} s_1 & & \\ & \ddots & \\ & & s_k \end{pmatrix} \in S$ with the Chern roots of the tautological bundle over X, and the Chern classes of the tautological line bundles over X_S. At fixed points of X_S, we have

$$s_a = \mathfrak{a}_{i_a}^{-1} \qquad a = 1, \cdots, k.$$

Then corresponding to (9.3.9) the stable envelope for X_S is given by the product of those for $T^*\mathbb{P}(\mathbb{C}^n)$ (9.3.8) with shifts in the Kähler parameters by

$$2\rho = (n-1, n-3, \cdots, 1-n),$$

(See 4.4.4 in [3]). Hence one obtains

$$\begin{aligned} \mathrm{Stab}_{\mathfrak{C}}^{X_S}(F_I) &= \prod_{a=1}^{k} \mathrm{Stab}_{\mathfrak{C}}^{T^*\mathbb{P}(\mathbb{C}^n)}(F_{i_a})|_{\zeta \mapsto \zeta \hbar^{2\rho_a}} \\ &= \prod_{a=1}^{k} \left[\prod_{1 \le i < i_a} \theta(s_a \mathfrak{a}_i) \frac{\theta(s_a \mathfrak{a}_{i_a} \zeta \hbar^{i_a - n + 2\rho_a})}{\theta(\zeta \hbar^{i_a - n + 2\rho_a})} \prod_{i_a < i \le n} \theta(s_a \mathfrak{a}_i \hbar) \right]. \end{aligned} \tag{9.3.10}$$

Note that

$$i_a - n + 2\rho_a = i_a - 2a + 1.$$

Then the abelianization formula [3, 146] gives the stable envelopes $\mathrm{Stab}_{\mathfrak{C}}(F_{I_1})$ for X in terms of $\mathrm{Stab}_{\mathfrak{C}}^{X_S}(F_I)$ as follows.

$$\mathrm{Stab}_{\mathfrak{C}}(F_I) = \mathrm{Sym}_{s_1, \cdots, s_k} \frac{\mathrm{Stab}_{\mathfrak{C}}^{X_S}(F_I)}{\prod_{1 \le a < b \le k} \theta(s_a/s_b)\theta(\hbar s_b/s_a)}. \tag{9.3.11}$$

9.4 Direct Comparison with the Elliptic Weight Functions

Let $X = T^*\mathrm{Gr}(k,n)$ and $\lambda = (k, n-k)$. Now we compare the expression of $\mathrm{Stab}_{\mathfrak{C}}(F_I)$ in (9.3.11) with the elliptic weight functions $\mathscr{W}_I(t,z,\Pi)$ obtained in Chap. 6. We show that they coincide with each other under certain identification of the variables.

First of all, the symmetry structure in the target of (9.2.1) coincides with the one of the elliptic weight function $\mathscr{W}_I(t,z,\Pi)$ in the variables t_a $(a = 1, \cdots, \lambda_1)$. This suggests that t_a can be identified with the Chern roots s_a of the tautological vector bundles over X [16]. This structure as well as the quasi-periodicity in Proposition 6.3.5 allows us to regard the elliptic weight functions $\mathscr{W}_I(t,z,\Pi)$ as meromorphic sections of line bundles over $\mathrm{E}_T(X)$ near the origin of $\mathscr{B}_{T,X}$.

More precisely, let us consider the elliptic weight functions

$$\mathscr{W}_{\sigma_0(I)}(\sigma_0^{(k)}(t), \sigma_0(z^{-1}), \Pi^{-1}),$$

where

$$\sigma_0(I) = (\tilde{I}_1, \tilde{I}_2), \qquad \tilde{I}_1 = \{\tilde{i}_1 < \tilde{i}_2 < \cdots < \tilde{i}_k\}, \qquad \tilde{i}_a = \sigma_0(i_{\sigma_0^{(k)}(a)}),$$

$$\sigma_0^{(k)}(t) = (s_{\sigma_0^{(k)}(a)}) \qquad (a = 1, \cdots, k),$$

$$\sigma_0(z^{-1}) = (z^{-1}_{\sigma_0(i)}) \qquad (i = 1, \cdots, n).$$

Set $\tilde{i} = \sigma_0(i) = n - i + 1$, $\tilde{a} = \sigma_0^{(k)}(a) = k - a + 1$ and note that

$$\tilde{i} \lesseqgtr \tilde{i}_{\tilde{a}} \quad \Leftrightarrow \quad i \gtreqless i_a.$$

In particular, for $C(i_a)$ in Proposition 6.2.2 one gets

$$-C(i_a) = -2(k-a) + n - i_a = \tilde{i}_{\tilde{a}} - 2\tilde{a} + 1.$$

Hence one finds

$$\begin{aligned}
&\mathscr{W}_{\sigma_0(I)}(\sigma_0^{(k)}(t), \sigma_0(z^{-1}), \Pi^{-1}) \\
&= (-1)^{k(n-1)} \mathrm{Sym}_{t_1, \cdots, t_k} \frac{\prod_{a=1}^{k} \left[\prod_{1 \le \tilde{i} < \tilde{i}_{\tilde{a}}} \theta(t_{\tilde{a}} z_{\tilde{i}}) \frac{\theta(t_{\tilde{a}} z_{\tilde{i}_{\tilde{a}}} \Pi q^{-2(\tilde{i}_{\tilde{a}} - 2\tilde{a} + 1)})}{\theta(\Pi q^{-2(\tilde{i}_{\tilde{a}} - 2\tilde{a} + 1)})} \prod_{\tilde{i}_{\tilde{a}} < \tilde{i} \le n} \theta(t_{\tilde{a}} z_{\tilde{i}} q^{-2}) \right]}{\prod_{1 \le \tilde{a} < \tilde{b} \le k} \theta(t_{\tilde{a}}/t_{\tilde{b}}) \theta(q^{-2} t_{\tilde{b}}/t_{\tilde{a}})}.
\end{aligned}$$

Comparing this with (9.3.10) and (9.3.11), we obtain

$$\mathrm{Stab}_{\mathfrak{C}}(F_I) = \mathscr{W}_{\sigma_0(I)}(\sigma_0^{(k)}(t), \sigma_0(z^{-1}), \Pi^{-1}) \tag{9.4.1}$$

under the identification

$$\hbar = q^{-2}, \tag{9.4.2}$$

$$s_a = t_{\tilde{a}} \qquad (a = 1, \cdots, k), \tag{9.4.3}$$

$$\mathfrak{a}_i = z_{\tilde{i}} \qquad (i = 1, \cdots, n), \tag{9.4.4}$$

$$\zeta = \Pi. \tag{9.4.5}$$

9.4.1 Restriction to the Fixed Points

For $J = (J_1, J_2) \in \mathscr{I}_\lambda$, let us consider the restriction of $\mathrm{Stab}_{\mathfrak{C}}(F_I)$ to the fixed point component F_J, i.e. the specialization $s_a = \mathfrak{a}_{j_a}^{-1}$, $(a = 1, \cdots, k)$. From (9.3.11) one obtains

$$\mathrm{Stab}_{\mathfrak{C}}(F_I)|_{F_J} = \mathrm{Sym}_s \frac{\prod_{a=1}^{k} f(\mathfrak{a}_{j_a}^{-1}, \mathfrak{a}, \zeta\hbar^{i_a-2a+1})}{\prod_{1\le a<b\le k} \theta(\mathfrak{a}_{j_b}/\mathfrak{a}_{j_a})\theta(\mathfrak{a}_{j_a}/\mathfrak{a}_{j_b}/\hbar)},$$

where

$$f(\mathfrak{a}_{j_a}^{-1}, \mathfrak{a}, \zeta\hbar^{i_a-2a+1})$$
$$= \prod_{1\le i<i_a} \theta(\mathfrak{a}_i/\mathfrak{a}_{j_a}) \frac{\theta(\mathfrak{a}_{i_a}\zeta\hbar^{\tilde{i}_{\tilde{a}}-2\tilde{a}+1}/\mathfrak{a}_{j_a})}{\theta(\zeta\hbar^{\tilde{i}_{\tilde{a}}-2\tilde{a}+1})} \prod_{i_a<i\le n} \theta(\mathfrak{a}_i\hbar/\mathfrak{a}_{j_a}).$$

This is an elliptic and dynamical analogue of the formula in Theorem 5.2.1 in [146] for $X = T^*\mathrm{Gr}(k, n)$. Then from (9.4.1), one obtains the following identification.

$$\mathrm{Stab}_{\mathfrak{C}}(F_I)|_{F_J} = \mathscr{W}_{\sigma_0(I)}(z_{\sigma_0(J)}^{-1}, \sigma_0(z^{-1}), \Pi^{-1}). \tag{9.4.6}$$

9.5 The Stable Classes and the Fixed Point Classes

Let $X_k = T^*\mathrm{Gr}(k, n)$ and fix a chamber $\mathfrak{C}$ as above. By definition, the stable classes $\mathrm{Stab}_{\mathfrak{C}}(F_K)$ $(K \in \mathscr{I}_\lambda)$ are triangular with respect to the fixed point classes $\{[I]\}_{I\in\mathscr{I}_\lambda}$ in $\mathrm{E}_T(X_k)$. Namely, one has the following expansion formula

$$\mathrm{Stab}_{\mathfrak{C}}(F_K) = \sum_{I\in\mathscr{I}_\lambda} \frac{\mathrm{Stab}_{\mathfrak{C}}(F_K)|_{F_I}}{R(z_I^{-1})} [I]. \tag{9.5.1}$$

Here we chose a normalization $R(z_I)$ given in Proposition 6.3.4 for later convenience. We regard this as the definition of the fixed point classes.

Note that from (9.4.6), one has

$$\mathrm{Stab}_{\mathfrak{C}}(F_K)|_{F_I} = \mathscr{W}_{\sigma_0(K)}(z^{-1}_{\sigma_0(I)}, \sigma_0(z^{-1}), \Pi^{-1}). \tag{9.5.2}$$

Note also that by the replacement $z \mapsto z^{-1}$ and $\Pi \mapsto \Pi^{-1}$ one can rewrite Proposition 6.3.4 as

$$\sum_{I \in \mathscr{I}_\lambda} \frac{\mathscr{W}_J(z_I^{-1}, z^{-1}, \Pi q^{2\sum_{j=1}^n \langle \bar{\epsilon}_{\mu_j}, h\rangle}) \mathscr{W}_{\sigma_0(K)}(z_I^{-1}, \sigma_0(z^{-1}), \Pi^{-1})}{Q(z_I^{-1}) R(z_I^{-1})} = \delta_{J,K}.$$

Then using this and (9.5.2) one can invert (9.5.1) and obtain

$$[I] = \sum_{J \in \mathscr{I}_\lambda} \widetilde{W}_J(z^{-1}_{\sigma_0(I)}, z^{-1}, \Pi q^{2\sum_{j=1}^n \langle \bar{\epsilon}_{\mu_j}, h\rangle})\, \mathrm{Stab}_{\mathfrak{C}}(F_J). \tag{9.5.3}$$

Comparing this with Theorem 7.2.3, one finds that (9.5.3) is identical to the relation (7.2.2) under the following correspondence

$$\begin{aligned} \text{the Gelfand-Tsetlin base } \xi_{\sigma_0(I)} &\Leftrightarrow \text{the fixed point class } [I], \\ \text{the standard base } v_J &\Leftrightarrow \text{the stable class } \mathrm{Stab}_{\mathfrak{C}}(F_J). \end{aligned} \tag{9.5.4}$$

Remark 9.2 Similar correspondences between the Gelfand-Tsetlin basis and the fixed point classes were studied in [39, 40, 89, 123, 154] for different algebras and geometries. In [123] for the level-(0,1) representation of the quantum toroidal algebra of type A, the Gelfand-Tsetlin basis on the q-Fock space [155] was identified with the fixed point basis of the equivariant K-theory of corresponding cyclic quiver variety [157]. Affine Yangian analogue of this result was obtained in [89]. In [39, 40, 154], certain geometric actions of the universal enveloping algebra $U(\mathfrak{gl}_N)$ on the Laumon spaces, of the affine Yangian of type $A^{(1)}_{N-1}$ on the affine Laumon spaces and of the quantum toroidal algebra $\ddot{U}_q(\widehat{\mathfrak{sl}}_N)$ on the K-theory of the affine Laumon spaces were constructed, respectively.

9.6 Geometric Representation

The correspondence in (9.5.4) indicates the isomorphism between $\widehat{V}_{z_1} \widetilde{\otimes} \cdots \widetilde{\otimes} \widehat{V}_{z_n}$ and the space $\mathrm{Span}_{\mathbb{C}}\{\mathrm{Stab}_{\mathfrak{C}}(F_I), I \in \mathscr{I}_{(k,n-k)}, 0 \le k \le n\}$ and suggests that the finite dimensional representation of $U_{q,p}(\widehat{\mathfrak{sl}}_2)$ can be constructed geometrically on $\bigoplus_{k=0}^{n} \mathrm{E}_T(X_k)$ in a similar way to [64, 126] by using "correspondences"[24]. To formulate such construction is an outstanding problem and beyond the scope of this book. Instead, we here present a heuristic argument based on the results in Theorem 7.3.1 and Corollary 7.3.2. Namely, one finds that all the formulas there in Sect. 7.3 are given in purely combinatorial way in terms of the partition of $[1, n]$. Hence by replacing the Gelfand-Tsetlin bases ξ_I with the fixed point classes $[I]$, one obtains a level-0 action of $U_{q,p}(\widehat{\mathfrak{sl}}_2)$ on $\bigoplus_{k=0}^{n} \mathrm{E}_T(X_k)$ from (9.5.3) and (9.5.4).

Theorem 9.6.1 *Under the same notation as Theorem 7.3.1, let us define the action of the half currents* $\widetilde{k}_j^{\pm}(w)$, $e^{\pm}(w,\zeta)$, $f^{\pm}(w,\zeta)$ *on the fixed point classes by*

$$\widetilde{k}_1^{\pm}(w)[I] = \prod_{b \in I_2} \left.\frac{\theta(\hbar w/\mathfrak{a}_b)}{\theta(w/\mathfrak{a}_b)}\right|_{\pm} [I], \quad \widetilde{k}_2^{\pm}(w)[I] = \prod_{a \in I_1} \left.\frac{\theta(w/\mathfrak{a}_b)}{\theta(\hbar^{-1} w/\mathfrak{a}_b)}\right|_{\pm} [I], \tag{9.6.1}$$

$$e^{\pm}(w,\zeta)[I] = \sum_{i \in I_2} \left.\frac{\theta(\zeta \mathfrak{a}_i/w)\theta(\hbar^{-1})}{\theta(\zeta)\theta(w/\mathfrak{a}_i)}\right|_{\pm} \prod_{\substack{k \in I_2 \\ \neq i}} \frac{\theta(\hbar^{-1}\mathfrak{a}_k/\mathfrak{a}_i)}{\theta(\mathfrak{a}_k/\mathfrak{a}_i)} [I^{i'}], \tag{9.6.2}$$

$$f^{\pm}(w,\zeta)[I] = \sum_{i \in I_1} \left.\frac{\theta(\zeta \hbar^{-(2k-n-1)} w/\mathfrak{a}_i)\theta(\hbar^{-1})}{\theta(\zeta \hbar^{-(2k-n-1)})\theta(w/\mathfrak{a}_i)}\right|_{\pm} \prod_{\substack{k \in I_1 \\ \neq i}} \frac{\theta(\hbar^{-1}\mathfrak{a}_i/\mathfrak{a}_k)}{\theta(\mathfrak{a}_i/\mathfrak{a}_k)} [I'^{i}]. \tag{9.6.3}$$

Then this gives an irreducible finite dimensional representation of $U_{q,p}(\widehat{\mathfrak{sl}}_2)$ *on* $\bigoplus_{k=0}^{n} \mathrm{E}_T(X_k)$.

Corollary 9.6.2 *The level-0 action of* $U_{q,p}(\widehat{\mathfrak{sl}}_2)$ *on* $\bigoplus_{k=0}^{n} \mathrm{E}_T(X_k)$ *is given by*

$$\psi^{\pm}(w)[I] = \varsigma \prod_{a \in I_1} \left.\frac{\theta(\hbar^{-1} w/\mathfrak{a}_a)}{\theta(w/\mathfrak{a}_a)}\right|_{\pm} \prod_{b \in I_2} \left.\frac{\theta(\hbar w/\mathfrak{a}_b)}{\theta(w/\mathfrak{a}_b)}\right|_{\pm} e^{-Q}[I],$$

$$e(w)[I] = -\frac{a^*\theta(\hbar^{-1})}{(p;p)_\infty^3} \sum_{i \in I_2} \delta(\mathfrak{a}_i/w) \prod_{\substack{k \in I_2 \\ \neq i}} \frac{\theta(\hbar^{-1}\mathfrak{a}_k/\mathfrak{a}_i)}{\theta(\mathfrak{a}_k/\mathfrak{a}_i)} e^{-Q}[I^{i'}],$$

$$f(w)[I] = -\frac{a\theta(\hbar^{-1})}{(p;p)_\infty^3} \sum_{i \in I_1} \delta(\mathfrak{a}_i/w) \prod_{\substack{k \in I_1 \\ \neq i}} \frac{\theta(\hbar^{-1}\mathfrak{a}_i/\mathfrak{a}_k)}{\theta(\mathfrak{a}_i/\mathfrak{a}_k)} [I'^{i}].$$

Appendix A
Affine Quantum Group $U_q(\mathfrak{g})$ in the Drinfeld–Jimbo Formulation

This appendix provides a summary of the Drinfeld–Jimbo formulation of the affine quantum group $U_q(\mathfrak{g})$ and associated universal R-matrix.

A.1 The Drinfeld–Jimbo Formulation

Let $\mathfrak{g}$ be the Kac–Moody Lie algebra with a symmetrizable generalized Cartan matrix $(a_{ij})_{i,j\in I}$. Let $(\ ,\ )$ denote an invariant inner product on the Cartan subalgebra $\mathfrak{h}$ and identify $\mathfrak{h}^*$ with $\mathfrak{h}$. For simple roots α_i, set $(\alpha_i,\alpha_j) = d_i a_{ij}$ with $d_i = (\alpha_i,\alpha_i)/2$. We choose the ground ring $\mathbb{C}[[\hbar]]$ with $q = e^\hbar$. The algebra $U = U_q(\mathfrak{g})$ is generated by e_i, f_i $(i \in I)$ and h $(h \in \mathfrak{h})$ subject to the relations

$$[h, h'] = 0 \qquad (h, h' \in \mathfrak{h}), \tag{A.1.1}$$

$$[h, e_i] = (h, \alpha_i)e_i, \qquad [h, f_i] = -(h, \alpha_i)f_i \qquad (i \in I, h \in \mathfrak{h}), \tag{A.1.2}$$

$$[e_i, f_j] = \delta_{ij}\frac{t_i - t_i^{-1}}{q_i - q_i^{-1}} \qquad (i, j \in I), \tag{A.1.3}$$

and the Serre relations which we omit. In (A.1.3) we have set $q_i = q^{d_i}$, $t_i = q^{\alpha_i}$. We adopt the Hopf algebra structure given as follows:

$$\Delta(h) = h \otimes 1 + 1 \otimes h, \tag{A.1.4}$$

$$\Delta(e_i) = e_i \otimes 1 + t_i \otimes e_i, \quad \Delta(f_i) = f_i \otimes t_i^{-1} + 1 \otimes f_i, \tag{A.1.5}$$

$$\varepsilon(e_i) = \varepsilon(f_i) = \varepsilon(h) = 0, \tag{A.1.6}$$

$$S(e_i) = -t_i^{-1}e_i, \quad S(f_i) = -f_i t_i, \quad S(h) = -h, \tag{A.1.7}$$

where $i \in I$ and $h \in \mathfrak{h}$.

H. Konno, *Elliptic Quantum Groups*, SpringerBriefs in Mathematical Physics 37,
https://doi.org/10.1007/978-981-15-7387-3

Let $\mathscr{R} \in U^{\otimes 2}$ denote the universal R matrix of U. It has the form [29, 152]

$$\mathscr{R} = q^{-T}\mathscr{C}, \tag{A.1.8}$$

$$\begin{aligned}\mathscr{C} &= \sum_{\beta \in Q^+} q^{(\beta,\beta)}\left(q^{-\beta} \otimes q^{\beta}\right)\mathscr{C}_\beta \\ &= 1 - \sum_{i \in I}(q_i - q_i^{-1})e_i t_i^{-1} \otimes t_i f_i + \cdots .\end{aligned} \tag{A.1.9}$$

Here

$$T = \sum_l h_l \otimes h^l, \tag{A.1.10}$$

with a basis $\{h_l\}$ of $\mathfrak{h}$ and its dual basis $\{h^l\}$, denotes the canonical element of $\mathfrak{h} \otimes \mathfrak{h}$. The element $\mathscr{C}_\beta = \sum_j u_{\beta,j} \otimes u^j_{-\beta}$ is the canonical element of $U^+_\beta \otimes U^-_{-\beta}$ with respect to a certain Hopf pairing, where U^+ (resp. U^-) denotes the subalgebra of U generated by the e_i (resp. f_i), and $U^\pm_{\pm\beta}$ ($\beta \in Q^+$) signifies the homogeneous components with respect to the natural gradation by $Q^+ = \sum_i \mathbb{Z}_{\geq 0}\alpha_i$.

The universal R matrix has the following properties:

$$\Delta'(a) = \mathscr{R}\Delta(a)\mathscr{R}^{-1} \qquad \forall a \in U, \tag{A.1.11}$$

$$(\Delta \otimes \mathrm{id})\,\mathscr{R} = \mathscr{R}^{(13)}\mathscr{R}^{(23)}, \tag{A.1.12}$$

$$(\mathrm{id} \otimes \Delta)\,\mathscr{R} = \mathscr{R}^{(13)}\mathscr{R}^{(12)}, \tag{A.1.13}$$

$$(\varepsilon \otimes \mathrm{id})\,\mathscr{R} = (\mathrm{id} \otimes \varepsilon)\,\mathscr{R} = 1. \tag{A.1.14}$$

Here $\Delta' = \sigma \circ \Delta$ signifies the opposite coproduct, σ being the flip of the tensor components $\sigma(a \otimes b) = b \otimes a$. From (A.1.11)–(A.1.13) follows the Yang–Baxter equation

$$\mathscr{R}^{(12)}\mathscr{R}^{(13)}\mathscr{R}^{(23)} = \mathscr{R}^{(23)}\mathscr{R}^{(13)}\mathscr{R}^{(12)}. \tag{A.1.15}$$

Appendix B
Elliptic Quantum Algebra $U_{q,p}(\widehat{\mathfrak{gl}}_2)$

In this appendix, we summarize a definition of the elliptic quantum algebra $U_{q,p}(\widehat{\mathfrak{gl}}_2)$.

B.1 Definition of $U_{q,p}(\widehat{\mathfrak{gl}}_2)$

Definition B.1.1 The elliptic algebra $U_{q,p}(\widehat{\mathfrak{gl}}_2)$ is a topological algebra over $\mathbb{F}[[p]]$ generated by $e_m, f_m, k_{l,m}, (l = 1, 2, m \in \mathbb{Z}), \widehat{d}$ and the central element $q^{c/2}$. We set

$$e(z) = \sum_{m\in\mathbb{Z}} e_m z^{-m}, \quad f(z) = \sum_{m\in\mathbb{Z}} f_m z^{-m}, \tag{B.1.1}$$

$$k_l^+(z) = \sum_{m\in\mathbb{Z}_{\geq 0}} k_{l,-m} z^m + \sum_{m\in\mathbb{Z}_{>0}} k_{l,m} p^m z^{-m}. \tag{B.1.2}$$

The defining relations are as follows. For $g(P) \in \mathbb{F}$,

$$\begin{aligned}&g(P)e(z) = e(z)g(P-2),\ g(P)f(z) = f(z)g(P),\\ &g(P)k_l^+(z) = k_l^+(z)g(P - \langle Q_{\bar{\epsilon}_l}, P\rangle),\ [g(P), \widehat{d}] = 0,\end{aligned} \tag{B.1.3}$$

$$[\widehat{d}, k_l^+(z)] = -z\frac{\partial}{\partial z}k_l^+(z),\ [\widehat{d}, e(z)] = -z\frac{\partial}{\partial z}e(z),\ [\widehat{d}, f(z)] = -z\frac{\partial}{\partial z}f(z), \tag{B.1.4}$$

$$\rho_+^+(z_2/z_1)k_l^+(z_1)k_l^+(z_2) = \rho_+^+(z_1/z_2)k_l^+(z_2)k_l^+(z_1), \tag{B.1.5}$$

$$\begin{aligned}&\rho_+^+(z_2/z_1)\frac{(p^*z_2/z_1; p^*)_\infty(pq^2z_2/z_1; p)_\infty}{(p^*q^2z_2/z_1; p^*)_\infty(pz_2/z_1; p)_\infty}k_1^+(z_1)k_2^+(z_2)\\ &\quad = \rho_+^+(z_1/z_2)\frac{(q^{-2}z_1/z_2; p^*)_\infty(z_1/z_z; p)_\infty}{(z_1/z_2; p^*)_\infty(q^{-2}z_1/z_2; p)_\infty}k_2^+(z_2)k_1^+(z_1),\end{aligned} \tag{B.1.6}$$

H. Konno, *Elliptic Quantum Groups*, SpringerBriefs in Mathematical Physics 37,
https://doi.org/10.1007/978-981-15-7387-3

$$\frac{(p^*q^{c+1}z_2/z_1;p^*)_\infty}{(p^*q^{c-1}z_2/z_1;p^*)_\infty}k_1^+(z_1)e(z_2)=q^{-1}\frac{(q^{-c+1}z_1/z_2;p^*)_\infty}{(q^{-c-1}z_1/z_2;p^*)_\infty}e(z_2)k_1^+(z_1), \quad (B.1.7)$$

$$\frac{(p^*q^{c-3}z_2/z_1;p^*)_\infty}{(p^*q^{c-1}z_2/z_1;p^*)_\infty}k_2^+(z_1)e(z_2)=q\frac{(q^{-c+1}z_1/z_2;p^*)_\infty}{(q^{-c+3}z_1/z_2;p^*)_\infty}e(z_2)k_2^+(z_1), \quad (B.1.8)$$

$$\frac{(pq^{-1}z_2/z_1;p)_\infty}{(pqz_2/z_1;p)_\infty}k_1^+(z_1)f(z_2)=q\frac{(q^{-1}z_1/z_2;p)_\infty}{(qz_1/z_2;p)_\infty}f(z_2)k_1^+(z_1), \quad (B.1.9)$$

$$\frac{(pq^{-1}z_2/z_1;p)_\infty}{(pq^{-3}z_2/z_1;p)_\infty}k_2^+(z_1)f(z_2)=q^{-1}\frac{(q^3z_1/z_2;p)_\infty}{(qz_1/z_2;p)_\infty}f(z_2)k_2^+(z_1), \quad (B.1.10)$$

$$z_1\frac{(q^2z_2/z_1;p^*)_\infty}{(p^*q^{-2}z_2/z_1;p^*)_\infty}e(z_1)e(z_2)=-z_2\frac{(q^2z_1/z_2;p^*)_\infty}{(p^*q^{-2}z_1/z_2;p^*)_\infty}e(z_2)e(z_1), \quad (B.1.11)$$

$$z_1\frac{(q^{-2}z_2/z_1;p)_\infty}{(pq^2z_2/z_1;p)_\infty}f(z_1)f(z_2)=-z_2\frac{(q^{-2}z_1/z_2;p)_\infty}{(pq^2z_1/z_2;p)_\infty}f(z_2)f(z_1), \quad (B.1.12)$$

$$[e(z_1),f(z_2)]=\frac{\varsigma}{q-q^{-1}}\Big(\delta(q^{-c}z_1/z_2)k_1^-(q^{-\frac{c}{2}}z_1)k_2^-(q^{-\frac{c}{2}}z_1)^{-1}$$
$$-\delta(q^{c}z_1/z_2)k_1^+(q^{-\frac{c}{2}}z_2)k_2^+(q^{-\frac{c}{2}}z_2)^{-1}\Big), \quad (B.1.13)$$

where $\delta(z)=\sum_{n\in\mathbb{Z}}z^n$, ς is given in Proposition 2.5.1, and

$$\rho_+^+(z)=\frac{(q^2z;p^*,q^4)_\infty^2}{(z;p^*,q^4)_\infty(q^4z;p^*,q^4)_\infty}\frac{(z;p,q^4)_\infty(q^4z;p,q^4)_\infty}{(q^2z;p,q^4)_\infty^2}. \quad (B.1.14)$$

We call $e(z)$, $f(z)$, $k_l^\pm(z)$ the elliptic currents. We also denote by $U'_{q,p}(\widehat{\mathfrak{gl}}_2)$ the subalgebra obtained by removing $\widehat{d}$.

We treat these relations as formal Laurent series in z, w, and z_j's. All the coefficients in z_j's are well defined in the p-adic topology.

Proposition B.1.1 *Relations* (B.1.5)–(B.1.12) *can be rewritten as follows:*

$$k_l^+(z_1)k_l^+(z_2)=\rho(z_1/z_2)k_l^+(z_2)k_l^+(z_1),, \quad (B.1.15)$$

$$k_1^+(z_1)k_2^+(z_2)=\rho(z_1/z_2)\frac{\theta^*(q^{-2}z_1/z_2)\theta(z_1/z_2)}{\theta^*(z_1/z_2)\theta(q^{-2}z_1/z_2)}k_2^+(z_2)k_1^+(z_1), \quad (B.1.16)$$

$$k_1^+(z_1)e(z_2)k_1^+(z_1)^{-1}=\frac{\theta^*(q^{-c+1}z_1/z_2)}{\theta^*(q^{-c-1}z_1/z_2)}e(z_2), \quad (B.1.17)$$

$$k_2^+(z_1)e_1(z_2)k_2^+(z_1)^{-1}=\frac{\theta^*(q^{-c+1}z_1/z_2)}{\theta^*(q^{-c+3}z_1/z_2)}e(z_2), \quad (B.1.18)$$

$$k_1^+(z_1)f(z_2)k_1^+(z_1)^{-1}=\frac{\theta(q^{-1}z_1/z_2)}{\theta(qz_1/z_2)}f(z_2), \quad (B.1.19)$$

$$k_2^+(z_1)f(z_2)k_2^+(z_1)^{-1} = \frac{\theta(q^3z_1/z_2)}{\theta(qz_1/z_2)}f(z_2), \tag{B.1.20}$$

$$e(z_1)e(z_2) = \frac{\theta^*(q^2z_1/z_2)}{\theta^*(q^{-2}z_1/z_2)}e(z_2)e(z_1), \tag{B.1.21}$$

$$f(z_1)f(z_2) = \frac{\theta(q^{-2}z_1/z_2)}{\theta(q^2z_1/z_2)}f(z_2)f(z_1), \tag{B.1.22}$$

Proposition B.1.2 *Let us set*

$$K(z) = k_1^+(z)k_2^+(q^{-2}z).$$

Then $K(z)$ belongs to the center of $U'_{q,p}(\widehat{\mathfrak{gl}}_2)$.

Proof Direct calculation using (B.1.3), (B.1.15)–(B.1.20) shows that $K(z)$ commutes with $\mathbb{F}$ and all of the elliptic currents of $U'_{q,p}(\widehat{\mathfrak{gl}}_2)$. In particular, $[K(z), k_l^+(w)] = 0$ $(l = 1, 2)$ follows from the identity

$$\rho(z)\rho(q^{-2}z) = \frac{\theta^*(z)\theta(q^{-1}z)}{\theta^*(q^{-1}z)\theta(z)}. \qquad \square$$

Remark B.1 $K(z)$ can be identified with the q-determinant of the L^+-operator.

The elliptic algebra $U'_{q,p}(\widehat{\mathfrak{sl}}_2)$ is the quotient algebra $U'_{q,p}(\widehat{\mathfrak{gl}}_2)/ < K(z) - 1 >$.

Appendix C
Central Extension of Felder's Elliptic Quantum Group $E_{q,p}(\widehat{\mathfrak{gl}}_2)$

This appendix provides a definition of the central extension of Felder's elliptic quantum group $E_{q,p}(\widehat{\mathfrak{gl}}_2)$. A brief discussion on the isomorphism between $U_{q,p}(\widehat{\mathfrak{gl}}_2)$ and $E_{q,p}(\widehat{\mathfrak{gl}}_2)$ is also given.

C.1 Definition of $E_{q,p}(\widehat{\mathfrak{gl}}_2)$

Let $\bar{L}_{ij,n}$ $(n \in \mathbb{Z}, 1 \leq i, j \leq 2)$ be abstract symbols. We define $L^+(z) = \sum_{1\leq i,j\leq 2} E_{ij} L^+_{ij}(z)$ by

$$L^+_{ij}(z) = \sum_{n\in\mathbb{Z}} L_{ij,n} z^{-n}, \qquad L_{ij,n} = p^{\max(n,0)} \bar{L}_{ij,n}. \tag{C.1.1}$$

Definition C.1.1 Let $R^+(z, \Pi)$ be the same R matrix as in Sect. 2.4. The elliptic algebra $E_{q,p}(\widehat{\mathfrak{gl}}_2)$ is a topological algebra over $\mathbb{F}[[p]]$ generated by $\bar{L}_{ij,n}$ $(i, j = 1, 2)$, $\widehat{d}$ and the central element $q^{c/2}$ satisfying the following relations:

$$R^{+(12)}(z_1/z_2, \Pi)L^{+(1)}(z_1)L^{+(2)}(z_2) = L^{+(2)}(z_2)L^{+(1)}(z_1)R^{+*(12)}(z_1/z_2, \Pi^*), \tag{C.1.2}$$

$$g(P)\bar{L}_{ij,n} = \bar{L}_{ij,n}\, g(P - \langle Q_{\bar{\epsilon}_j}, P\rangle), \tag{C.1.3}$$

$$[\widehat{d}, L^+(z)] = -z\frac{\partial}{\partial z}L^+(z), \ [\widehat{d}, g(P)] = 0, \tag{C.1.4}$$

where $g(P) \in \mathbb{F}$ and

$$L^{+(1)}(z) = L^+(z) \otimes \mathrm{id}, \qquad L^{+(2)}(z) = \mathrm{id} \otimes L^+(z).$$

H. Konno, *Elliptic Quantum Groups*, SpringerBriefs in Mathematical Physics 37,
https://doi.org/10.1007/978-981-15-7387-3

We regard $L^+(z) \in \mathrm{End}V \otimes E_{q,p}(\widehat{\mathfrak{gl}}_2)$. We treat (1.2.1) as a formal Laurent series in z_1 and z_2. Then the coefficients of z_1, z_2 are well defined in the p-adic topology. Note also that due to the RLL-relation (1.2.1) the L-operator $L^+(z)$ is invertible.

We also define $L^-(z) = \sum_{i,j=1,2} E_{ij} L^-_{ij}(z)$ in the same way as in Sect. 2.6

$$L^-(z) = \left(\mathrm{Ad}(q^{2\theta_V(P)}) \otimes \mathrm{id}\right)\left(q^{2T_V} L^+(zp^*q^c)\right). \qquad \text{(C.1.5)}$$

Then $L^\pm(z)$ satisfy the following RLL-relations derived in the same way as Proposition 2.6.1.

$$R^{\pm(12)}(z_1/z_2, \Pi)L^{\pm(1)}(z_1)L^{\pm(2)}(z_2) = L^{\pm(2)}(z_2)L^{\pm(1)}(z_1)R^{\pm*(12)}(z_1/z_2, \Pi^*), \qquad \text{(C.1.6)}$$

$$R^{\pm(12)}(q^{\pm k}z_1/z_2, \Pi)L^{\pm(1)}(z_1)L^{\mp(2)}(z_2) = L^{\mp(2)}(z_2)L^{\pm(1)}(z_1)R^{\pm*(12)}(q^{\mp k}z_1/z_2, \Pi^*). \qquad \text{(C.1.7)}$$

Remark C.1 One can expand (C.1.6) in both $z = z_1/z_2$ and $z^{-1} = z_2/z_1$. However (C.1.7) admits an expansion only in z (resp. z^{-1}) for the upper (resp. lower) sign case for the sake of the well-definedness in the p-adic topology. It is instructive to compare this with the trigonometric case [28].

C.2 The Gauss Coordinates of the $L^\pm$ Operators

The Gauss coordinates of the $L^\pm$ operator of $\mathcal{E} = E_{q,p}(\widehat{\mathfrak{gl}}_2)$ are introduced in the following way. Let us set

$$\mathcal{E}^\pm = \left\{A(z) \in \mathcal{E}[[p]][[z, z^{-1}]] \;\middle|\; A(z) \in \mathcal{E}[[z^{\pm 1}]] \mod p\mathcal{E}[[p]][[z, z^{-1}]]\right\}.$$

Then it is easy to show

Lemma C.2.1 *For $A(z), B(z) \in \mathcal{E}^\pm$, the product $A(z)B(z)$ is a well-defined element in $\mathcal{E}^\pm$ in the p-adic topology, respectively. Conversely, if $A(z), B(z) \in \mathcal{E}[[p]][[z, z^{-1}]]$ satisfy $A(z)B(z) \in \mathcal{E}^\pm$, then $A(z), B(z) \in \mathcal{E}^\pm$, respectively.*

Definition C.2.1 We define the Gauss coordinates $E^\pm(z), F^\pm(z), K_l^\pm(z)$ ($l = 1, 2$) of the L-operator $L^\pm(z)$ as follows:

$$L^\pm(z) = \begin{pmatrix} 1 & F^\pm(z) \\ 0 & 1 \end{pmatrix}\begin{pmatrix} K_1^\pm(z) & 0 \\ 0 & K_2^\pm(z) \end{pmatrix}\begin{pmatrix} 1 & 0 \\ E^\pm(z) & 1 \end{pmatrix}. \qquad \text{(C.2.1)}$$

Remark C.2 The Gauss coordinates $E^{\pm}(z), F^{\pm}(z), K_l^{\pm}(z)$ $(l = 1, 2)$ can be expressed in terms of the elliptic quantum minor determinants of $L^{\pm}(z)$. See Appendix E in [101].

Remark C.3 By definition the matrix elements $L_{ij}^{\pm}(z)$ are the elements in $\mathcal{E}^{\pm}$, respectively. Then from Lemma C.2.1, the matrix elements $E^{\pm}(z), F^{\pm}(z), K_l^{\pm}(z)$ of the right-hand side of (C.2.1) are elements in $\mathcal{E}^{\pm}$, respectively, and their products are well-defined formal Laurent series in z in the p-adic topology. In addition, since $L^{\pm}(z)$ are invertible, $K_l^{\pm}(z)$ $(1 \le l \le 2)$ are invertible. Therefore all the coordinates $E^{\pm}(z), F^{\pm}(z)$ and $K_l^{\pm}(z)$ are determined uniquely by $L_{ij}^{\pm}(z)$, respectively.

Hence we define the coefficients of the Gauss coordinates $E^{+}(z), F^{+}(z), K_l^{+}(z)$ as follows:

Definition C.2.2

$$E^{+}(z) = \sum_{n \in \mathbb{Z}_{\ge 0}} E_{-n}^{+} z^{n} + \sum_{n \in \mathbb{Z}_{>0}} E_{n}^{+} p^{n} z^{-n}, \tag{C.2.2}$$

$$F^{+}(z) = \sum_{n \in \mathbb{Z}_{\ge 0}} F_{-n}^{+} z^{n} + \sum_{n \in \mathbb{Z}_{>0}} F_{n}^{+} p^{n} z^{-n}, \tag{C.2.3}$$

$$K_j^{+}(z) = \sum_{n \in \mathbb{Z}_{\ge 0}} K_{j,-n}^{+} z^{n} + \sum_{n \in \mathbb{Z}_{>0}} K_{j,n}^{+} p^{n} z^{-n}. \tag{C.2.4}$$

We also remark that $E^{-}(z), F^{-}(z), K_l^{-}(z)$ are related to $E^{+}(z), F^{+}(z), K_l^{+}(z)$ in the same way as in (2.6.8)–(2.6.10).

C.3 Identification with the Elliptic Currents of $U_{q,p}(\widehat{\mathfrak{sl}}_2)$

Let $E_{q,p}(\widehat{\mathfrak{gl}}_2)_k = E_{q,p}(\widehat{\mathfrak{gl}}_2)/ < q^{c/2} - q^{k/2} >$. Let us consider the differences between the plus and the minus Gauss coordinates:

$$E(zq^{1-k/2}) := \mu^{*}(E^{+}(z^{+}) - E^{-}(z^{-})),$$
$$F(zq^{1-k/2}) := \mu(F^{+}(z^{-}) - F^{-}(z^{+})),$$

where we set $z^{\pm} = q^{\pm k/2} z$, and μ^{*}, μ are constants to be determined below. We show that these $E(z), F(z)$ as well as $K_l^{+}(z)$ $(l = 1, 2)$ satisfy the defining relations of $U_{q,p}(\widehat{\mathfrak{gl}}_2)_k$ and hence are identified with the elliptic currents of it. Combining this with Theorem 2.5.3 one shows the isomorphism between $U_{q,p}(\widehat{\mathfrak{gl}}_2)_k$ and $E_{q,p}(glth)_k$. A key to this is the following relations obtained from the RLL-relations (C.1.6)–(C.1.7) in the same way as in the proof of Theorem 2.5.3. For details see Appendix C in [101].

Proposition C.3.1 *Let* $z = z_1/z_2$.

$$K_l^{\pm}(z_1)K_l^{\pm}(z_2) = \rho(z)K_l^{\pm}(z_2)K_l^{\pm}(z_1), \tag{C.3.1}$$

$$K_l^{\pm}(z_1)K_l^{\mp}(z_2) = \frac{\rho^{\pm *}(zq^{\mp k})}{\rho^{\pm}(zq^{\pm k})}K_l^{\mp}(z_2)K_l^{\pm}(z_1), \tag{C.3.2}$$

$$K_1^{\pm}(z_2)K_2^{\pm}(z_1) = \rho(z)\frac{\bar{b}(z)}{\bar{b}^*(z)}K_2^{\pm}(z_1)K_1^{\pm}(z_2), \tag{C.3.3}$$

$$K_2^{\pm}(z_1)K_1^{\mp}(z_2) = \frac{\rho^{\pm *}(zq^{\mp k})}{\rho^{\pm}(zq^{\pm k})}\frac{\bar{b}^*(zq^{\mp k})}{\bar{b}(zq^{\pm k})}K_1^{\mp}(z_2)K_2^{\pm}(z_1), \tag{C.3.4}$$

$$K_2^{\pm}(z_1)^{-1}E^{\pm}(z_2)K_2^{\pm}(z_1) = E^{\pm}(z_2)\frac{1}{\bar{b}^*(z)} - E^{\pm}(z_1)\frac{c^*(z,\Pi^*)}{\bar{b}^*(z)}, \tag{C.3.5}$$

$$K_2^{\pm}(z_1)^{-1}E^{\mp}(z_2)K_2^{\pm}(z_1) = E^{\mp}(z_2)\frac{1}{\bar{b}^*(zq^{\mp k})} - E^{\pm}(z_1)\frac{c^*(zq^{\mp k},\Pi^*)}{\bar{b}^*(zq^{\mp k})}, \tag{C.3.6}$$

$$K_2^{\pm}(z_1)F^{\pm}(z_2)K_2^{\pm}(z_1)^{-1} = \frac{1}{\bar{b}(z)}F^{\pm}(z_2) - \frac{\bar{c}(z,\Pi)}{\bar{b}(z)}F^{\pm}(z_1), \tag{C.3.7}$$

$$K_2^{\pm}(z_1)F^{\mp}(z_2)K_2^{\pm}(z_1)^{-1} = \frac{1}{\bar{b}(zq^{\pm k})}F^{\mp}(z_2) - \frac{\bar{c}(zq^{\pm k},\Pi)}{\bar{b}(zq^{\pm k})}F^{\pm}(z_1), \tag{C.3.8}$$

$$K_1^{\pm}(z_2)^{-1}E^{\pm}(z_1)K_1^{\pm}(z_2) = \frac{1}{\bar{b}^*(z)}E^{\pm}(z_1) - \frac{\bar{c}^*(z,\Pi^*)}{\bar{b}^*(z)}E^{\pm}(z_2), \tag{C.3.9}$$

$$K_1^{\mp}(z_2)^{-1}E^{\pm}(z_1)K_1^{\mp}(z_2) = \frac{1}{\bar{b}^*(zq^{\mp k})}E^{\pm}(z_1) - \frac{\bar{c}^*(zq^{\mp k},\Pi^*)}{\bar{b}^*(zq^{\mp k})}E^{\mp}(z_2), \tag{C.3.10}$$

$$K_1^{\pm}(z_2)F^{\pm}(z_1)K_1^{\pm}(z_2)^{-1} = F^{\pm}(z_1)\frac{1}{\bar{b}(z)} - F^{\pm}(z_2)\frac{\bar{c}(z,\Pi)}{\bar{b}(z)}, \tag{C.3.11}$$

$$K_2^{\pm}(z_2)^{-1}E^{\pm}(z_1)K_2^{\pm}(z_2)E^{\pm}(z_2) = K_2^{\pm}(z_1)^{-1}E^{\pm}(z_2)K_2^{\pm}(z_1)E^{\pm}(z_1), \tag{C.3.12}$$

$$E^{\pm}(z_1)E^{\pm}(z_2)\frac{1}{\bar{b}^*(1/z)} - E^{\pm}(z_2)^2\frac{c^*(1/z,\Pi^*q^{-4})}{\bar{b}^*(1/z)}$$

$$= E^{\pm}(z_2)E^{\pm}(z_1)\frac{1}{\bar{b}^*(z)} - E^{\pm}(z_1)^2\frac{c^*(z,\Pi^*q^{-4})}{\bar{b}^*(z)}, \tag{C.3.13}$$

$$K_2^{\mp}(z_2)^{-1}E^{\pm}(z_1)K_2^{\mp}(z_2)E^{\mp}(z_2) = K_2^{\pm}(z_1)^{-1}E^{\mp}(z_2)K_2^{\pm}(z_1)E^{\pm}(z_1), \tag{C.3.14}$$

$$E^{\pm}(z_1)E^{\mp}(z_2)\frac{1}{\bar{b}^*(1/(zq^{\mp k}))} - E^{\mp}(z_2)^2\frac{c^*(1/(zq^{\mp k}), \Pi^* q^{-4})}{\bar{b}^*(1/(zq^{\mp k}))}$$
$$= E^{\mp}(z_2)E^{\pm}(z_1)\frac{1}{\bar{b}^*(zq^{\mp k})} - E^{\pm}(z_1)^2\frac{c^*(zq^{\mp k}, \Pi^* q^{-4})}{\bar{b}^*(zq^{\mp k})}, \quad \text{(C.3.15)}$$

$$F^{\pm}(z_1)F^{\pm}(z_2)\frac{1}{\bar{b}(z)} - F^{\pm}(z_1)^2\frac{\bar{c}(z, \Pi q^{-4})}{\bar{b}(z)}$$
$$= F^{\pm}(z_2)F^{\pm}(z_1)\frac{1}{\bar{b}(1/z)} - F^{\pm}(z_2)^2\frac{\bar{c}(1/z, \Pi q^{-4})}{\bar{b}(1/z)}, \quad \text{(C.3.16)}$$

$$F^{\pm}(z_1)F^{\mp}(z_2)\frac{1}{\bar{b}(zq^{\pm k})} + F^{\pm}(z_1)^2\frac{\bar{c}(zq^{\pm k}, \Pi q^{-4})}{\bar{b}(zq^{\pm k})}$$
$$= F^{\mp}(z_2)F^{\pm}(z_1)\frac{1}{\bar{b}(1/(zq^{\pm k}))} - F^{\pm}(z_1)^2\frac{\bar{c}(1/(zq^{\pm k}), \Pi q^{-4})}{\bar{b}(1/(zq^{\pm k}))}, \quad \text{(C.3.17)}$$

$$\rho^{\pm}(z)\left\{\bar{b}(z)K_2^{\pm}(z_1)E^{\pm}(z_1)F^{\pm}(z_2)K_2^{\pm}(z_2)\right.$$
$$\left.+\bar{c}(z,\Pi)\left(K_1^{\pm}(z_1) + F^{\pm}(z_1)K_2^{\pm}(z_1)E^{\pm}(z_1)\right)K_2^{\pm}(z_2)\right\}$$
$$= \rho^{*\pm}(z)\left\{\left(K_1^{\pm}(z_2) + F^{\pm}(z_2)K_2^{\pm}(z_2)E^{\pm}(z_2)\right)K_2^{\pm}(z_1)\bar{c}^*(z,\Pi^*)\right.$$
$$\left.+F^{\pm}(z_2)K_2^{\pm}(z_2)K_2^{\pm}(z_1)E^{\pm}(z_1)b^*(z,\Pi^*)\right\}, \quad \text{(C.3.18)}$$

$$[E^{\pm}(z_1), F^{\pm}(z_2)] = K_1^{\pm}(z_2)K_2^{\pm}(z_2)^{-1}\frac{\bar{c}^*(z, \Pi^* q^{-2})}{\bar{b}^*(z)}$$
$$- K_2^{\pm}(z_1)^{-1}K_1^{\pm}(z_1)\frac{\bar{c}(z, \Pi q^{-2})}{\bar{b}(z)}, \quad \text{(C.3.19)}$$

$$\rho^{\pm}(zq^{\pm k})\left(\bar{b}(zq^{\pm k})K_2^{\pm}(z_1)E^{\pm}(z_1)F^{\mp}(z_2)K_2^{\mp}(z_2)\right.$$
$$\left.+\bar{c}(zq^{\pm k},\Pi)(K_1^{\pm}(z_1) + F^{\pm}(z_1)K_2^{\pm}(z_1)E^{\pm}(z_1))K_2^{\mp}(z_2)\right)$$
$$= \rho^{*\pm}(zq^{\mp k})\left((K_1^{\mp}(z_2) + F^{\mp}(z_2)K_2^{\mp}(z_2)E^{\mp}(z_2))K_2^{\pm}(z_1)\bar{c}^*(zq^{\mp k},\Pi^*)\right.$$
$$\left.+F^{\mp}(z_2)K_2^{\mp}(z_2)K_2^{\pm}(z_1)E^{\pm}(z_1)b^*(zq^{\mp k},\Pi^*)\right), \quad \text{(C.3.20)}$$

$$[E^{\pm}(z_1), F^{\mp}(z_2)] = K_1^{\mp}(z_2)K_2^{\mp}(z_2)^{-1}\frac{\bar{c}^*(zq^{\mp k}, \Pi^* q^{-2})}{\bar{b}^*(zq^{\mp k})}$$
$$- K_2^{\pm}(z_1)^{-1}K_1^{\pm}(z_1)\frac{\bar{c}(zq^{\pm k}, \Pi q^{-2})}{\bar{b}(zq^{\pm k})}. \quad \text{(C.3.21)}$$

The identification with the elliptic currents is given as follows. Comparing (C.3.1), (C.3.3) with (B.1.15), (B.1.16), we identify $K_l^+(z)$ with $k_l^+(z)$ ($l = 1, 2$).

To obtain (B.1.18), we use (C.3.5) and (C.3.6) and derive

$$\begin{aligned}
K_2^+(z_1)^{-1}E(z_2q^{1-k/2})K_2^+(z_1) &= K_2^+(z_1)^{-1}\mu^*\left(E^+(z_2^+) - E^-(z_2^-)\right)K_2^+(z_1)\\
= \mu^*\Bigg(E^+(z_2^+)\frac{1}{\bar{b}^*(zq^{-k/2})} &- E^+(z_1)\frac{c^*(zq^{-k/2}, \Pi^*)}{\bar{b}^*(zq^{-k/2})}\\
-E^-(z_2^-)\frac{1}{\bar{b}^*(zq^{-k/2})} &+ E^+(z_1)\frac{c^*(zq^{-k/2}, \Pi^*)}{\bar{b}^*(zq^{-k/2})}\Bigg)\\
= E(z_2q^{1-k/2})\frac{1}{\bar{b}^*(zq^{-k/2})}. & \qquad\qquad (C.3.22)
\end{aligned}$$

Next, inserting (C.3.13) and (C.3.15) into $(E^+(z_1^+) - E^-(z_1^-))(E^+(z_2^+) - E^-(z_2^-))$, we obtain

$$E(z_1)E(z_2) = \frac{\theta^*(q^2z)}{\theta^*(q^{-2}z)}E(z_2)E(z_1).$$

This is identical to (B.1.21).

Similarly, we recover the relations (B.1.22) from (C.3.7)–(C.3.8), (C.3.16)–(C.3.17).

Finally let us check the relation (B.1.13). From (C.3.19) and (C.3.21), we have

$$\begin{aligned}
&(\mu\mu^*)^{-1}[E(z_1), F(z_2)]\\
&= [E^+(z_1^+) - E^-(z_1^-), F^+(z_2^-) - F^-(z_2^+)]\\
&= [E^+(z_1^+), F^+(z_2^-)] + [E^-(z_1^-), F^-(z_2^+)]\\
&\quad -[E^+(z_1^+), F^-(z_2^+)] - [E^-(z_1^-), F^+(z_2^-)]. \qquad (C.3.23)
\end{aligned}$$

Then substitute (C.3.19) and (C.3.21) into this. Noting Remark C.1, one finds that the terms containing $K_1^-(z_2^+)K_2^-(z_2^+)^{-1}$ from the 2nd and 3rd terms in (C.3.23) cancel out each other and the same is true for the terms containing $K_2^+(z_1^+)^{-1}K_1^+(z_1^+)$ from the 1st and the 3rd terms in (C.3.23). We thus obtain

$$\begin{aligned}
&(\mu\mu^*)^{-1}[E(z_1), F(z_2)]\\
&= -K_1^+(z_2^-)K_2^+(z_2^-)^{-1}\theta^*(q^2)\left(\left.\frac{\theta^*(\Pi^{*-1}q^2q^kz)}{\theta^*(\Pi^{*-1}q^2)\theta^*(zq^k)}\right|_+ - \left.\frac{\theta^*(\Pi^{*-1}q^2q^kz)}{\theta^*(\Pi^{*-1}q^2)\theta^*(zq^k)}\right|_-\right)\\
&\quad +K_2^-(z_1^-)^{-1}K_1^-(z_1^-)\theta(q^2)\left(\left.\frac{\theta(\Pi^{-1}q^2q^{-k}z)}{\theta(\Pi^{-1}q^2)\theta(zq^{-k})}\right|_+ - \left.\frac{\theta(\Pi^{-1}q^2q^{-k}z)}{\theta(\Pi^{-1}q^2)\theta(zq^{-k})}\right|_-\right).
\end{aligned}$$

These differences give the formal delta function $\delta(z) = \sum_{n\in\mathbb{Z}} z^n$ as in (4.2.8). We hence obtain (B.1.13) by making the constants μ^*, μ satisfy

$$\mu\mu^* \frac{\theta^*(q^2)}{(p^*;p^*)^3_\infty} = \frac{\varsigma}{q-q^{-1}}.$$

Appendix D
Proof of Theorem 7.2.3

In this appendix, we present the proof of Theorem 7.2.3.

D.1 Recursion Formula for X

Let $J = I_{\mu_1\cdots\mu_i\mu_{i+1}\cdots\mu_n} \in \mathscr{I}_\lambda$. By definition,

$$\begin{aligned}
\xi_{s_i(I)} &= \sum_J X_{s_i(I)J}(z,\Pi^*)v_J \\
&= \widetilde{S}_i(\Pi^*)\xi_I = \sum_J X_{IJ}(s_i(z),\Pi^*)\widetilde{S}_i(\Pi^*)v_J \\
&= \sum_{J,\mu_i',\mu_{i+1}'} X_{IJ}(s_i(z),\Pi^*)\overline{R}(z_i/z_{i+1},\Pi^* q^{2\sum_{j=1}^{i-1}\langle\bar{\epsilon}_{\mu_j},h\rangle})_{\mu_i'\mu_{i+1}'}^{\mu_i\mu_{i+1}} \\
&\qquad\qquad \times v_{\mu_1}\widetilde{\otimes}\cdots\widetilde{\otimes}v_{\mu_i'}\widetilde{\otimes}v_{\mu_{i+1}'}\widetilde{\otimes}\cdots\widetilde{\otimes}v_{\mu_n}.
\end{aligned}$$

Hence we obtain

$$X_{s_i(I)J}(z,\Pi^*) = X_{IJ}(s_i(z),\Pi^*) \tag{D.1.1}$$

for $\mu_i = \mu_{i+1}$, and

$$\begin{aligned}
&\left(X_{s_i(I)J}(z,\Pi^*)\ X_{s_i(I)s_i(J)}(z,\Pi^*)\right) \\
&= \left(X_{IJ}(s_i(z),\Pi^*)\ X_{Is_i(J)}(s_i(z),\Pi^*)\right) \mathsf{P}_2\ {}^t\overline{R}(z_i/z_{i+1},\Pi^* q^{2\sum_{j=1}^{i-1}\langle\bar{\epsilon}_{\mu_j},h\rangle})_{\mu_i,\mu_{i+1}}
\end{aligned} \tag{D.1.2}$$

H. Konno, *Elliptic Quantum Groups*, SpringerBriefs in Mathematical Physics 37,
https://doi.org/10.1007/978-981-15-7387-3

for $\mu_i > \mu_{i+1}$. Here we set

$$\mathsf{P}_2 = \begin{pmatrix} 0 & 1 \\ 1 & 0 \end{pmatrix}, \quad \overline{R}(z, \Pi^*)_{\mu_i, \mu_{i+1}} = \begin{pmatrix} \overline{R}(z, \Pi^*)^{\mu_{i+1}\mu_i}_{\mu_{i+1}\mu_i} & \overline{R}(z, \Pi^*)^{\mu_i\mu_{i+1}}_{\mu_{i+1}\mu_i} \\ \overline{R}(z, \Pi^*)^{\mu_{i+1}\mu_i}_{\mu_i\mu_{i+1}} & \overline{R}(z, \Pi^*)^{\mu_i\mu_{i+1}}_{\mu_i\mu_{i+1}} \end{pmatrix}. \tag{D.1.3}$$

Note that (D.1.1) and (D.1.2) determine the whole matrix elements in $\widehat{X}$ recursively starting from $X_{J^{max}J^{max}}(z, \Pi^*) = 1$.

On the other hand, from Proposition 6.3.2 with replacing Π by Π^* we have

$$\widetilde{W}_J(t, s_i(z), \Pi^*) = \widetilde{W}_J(t, z, \Pi^*) \tag{D.1.4}$$

if $\mu_i = \mu_{i+1}$, and

$$\begin{aligned} &\left(\widetilde{W}_J(t, s_i(z), \Pi^*)\ \widetilde{W}_{s_i(J)}(t, s_i(z), \Pi^*)\right) \\ &= \left(\widetilde{W}_J(t, z, \Pi^*)\ \widetilde{W}_{s_i(J)}(t, z, \Pi^*)\right) \mathsf{P}_2\, {}^t\overline{R}(z_i/z_{i+1}, \Pi^* q^{-2\sum_{j=i}^n \langle \bar{\epsilon}_{\mu_j}, h\rangle})_{\mu_i, \mu_{i+1}} \end{aligned}$$

if $\mu_i \neq \mu_{i+1}$. Using

$$\left(\mathsf{P}_2\, {}^t\overline{R}(z, \Pi^*)_{\mu_i, \mu_{i+1}}\right)^{-1} = \mathsf{P}_2\, {}^t\overline{R}(z^{-1}, \Pi^*)_{\mu_i, \mu_{i+1}},$$

we obtain in particular for $\mu_i > \mu_{i+1}$

$$\begin{aligned} &\left(\widetilde{W}_J(t, z, \Pi^*)\ \widetilde{W}_{s_i(J)}(t, z, \Pi^*)\right) \\ &= \left(\widetilde{W}_J(t, s_i(z), \Pi^*)\ \widetilde{W}_{s_i(J)}(t, s_i(z), \Pi^*)\right) \mathsf{P}_2\, {}^t\overline{R}(z_{i+1}/z_i, \Pi^* q^{-2\sum_{j=i}^n \langle \bar{\epsilon}_{\mu_j}, h\rangle})_{\mu_i, \mu_{i+1}} \end{aligned} \tag{D.1.5}$$

Specializing $t = s_i(z)_I$ and noting

$$\widetilde{W}_J(s_i(z)_I, z, \Pi^*) = \widetilde{W}_J(z_{s_i(I)}, z, \Pi^*)$$

etc., we obtain from (D.1.4) and (D.1.5)

$$\widetilde{W}_J(s_i(z)_I, s_i(z), \Pi^*) = \widetilde{W}_J(z_{s_i(I)}, z, \Pi^*) \tag{D.1.6}$$

if $\mu_i = \mu_{i+1}$, and

$$\left(\widetilde{W}_J(z_{s_i(I)}, z, \Pi^*)\ \widetilde{W}_{s_i(J)}(z_{s_i(I)}, z, \Pi^*)\right)$$
$$= \left(\widetilde{W}_J(s_i(z)_I, s_i(z), \Pi^*)\ \widetilde{W}_{s_i(J)}(s_i(z)_I, s_i(z), \Pi^*)\right) \mathsf{P}_2{}^t\overline{R}(z_{i+1}/z_i, \Pi^* q^{-2\sum_{j=i}^{n}\langle \bar{\epsilon}_{\mu_j}, h\rangle})_{\mu_i, \mu_{i+1}} \tag{D.1.7}$$

if $\mu_i > \mu_{i+1}$. Therefore one finds that $\widetilde{W}_J(z_I^{-1}, z^{-1}, \Pi^* q^{2\sum_{j=1}^{n}\langle \bar{\epsilon}_{\mu_j}, h\rangle})$ satisfy the same recursion relations as (D.1.1) and (D.1.2) for $X_{IJ}(z, \Pi^*)$. In addition their initial conditions coincide: $\widetilde{W}_{I^{max}}(z_{I^{max}}^{-1}, z^{-1}, \Pi^* q^{2\sum_{j=1}^{n}\langle \bar{\epsilon}_{\mu_j}, h\rangle}) = 1 = X_{I^{max} I^{max}}(z, \Pi^*)$.

Appendix E
Calculation of Trace

In this appendix, we summarize formulas for calculating a trace of operators on the Fock space $\mathscr{F}$ (4.3.2).

E.1 Coherent States

Let us consider the lebel-1 ($c = 1$) Heisenberg subalgebra (2.3.6)

$$[\alpha_m, \alpha_n] = \mathscr{C}_m \delta_{m+n,0},$$

where we set

$$\mathscr{C}_m = \frac{[2m]_q [m]_q}{m} \frac{1-p^m}{1-p^{*m}} q^{-m}.$$

We also need the scaling operator

$$\bar{d} = -\sum_{m>0} \frac{m}{\mathscr{C}_m} \alpha_{-m}\alpha_m.$$

Let $\xi_n, \bar{\xi}_n$ ($n \in \mathbb{Z}_{\neq 0}$) be complex conjugate variables and set $\xi = \{\xi_n\}$, etc. Define

$$|\xi\rangle = e^{\sum_{n>0} \frac{\xi_n \alpha_{-n}}{\mathscr{C}_n}} |0\rangle, \qquad \langle \bar{\xi}| = \langle 0| e^{\sum_{n>0} \frac{\bar{\xi}_n \alpha_n}{\mathscr{C}_n}}, \tag{E.1.1}$$

H. Konno, *Elliptic Quantum Groups*, SpringerBriefs in Mathematical Physics 37,
https://doi.org/10.1007/978-981-15-7387-3

where $|0\rangle$ and $\langle 0|$ are vacuum vectors

$$\alpha_n|0\rangle = 0, \qquad \langle 0|\alpha_{-n} = 0 \qquad (n \in \mathbb{Z}_{>0}).$$

Then we have

Proposition E.1.1

(1) $\alpha_n|\xi\rangle = \xi_n|\xi\rangle, \qquad \langle\bar{\xi}|\alpha_{-n} = \langle\bar{\xi}|\bar{\xi}_n$

(2) $\langle\bar{\xi}|\xi\rangle = \exp\left(\sum_{n>0} \frac{1}{\mathscr{C}_n}\bar{\xi}_n\xi_n\right)$

(3) $\langle\bar{\xi}|q^{\kappa\bar{d}} = \langle\{\bar{\xi}_n q^{\kappa n}\}|$

(4) *The coherent states* $\{|\xi\rangle\}$ *(resp.* $\{\langle\bar{\xi}|\}$*) form a complete basis of the Fock module* $\mathscr{F}$ *(resp.* $\mathscr{F}^*$*).*

$$id_{\mathscr{F}} = \int \prod_{n>0} \frac{d\xi_n\bar{\xi}_n}{\mathscr{C}_n} \exp\left(-\sum_{n>0} \frac{1}{\mathscr{C}_n}\bar{\xi}_n\xi_n\right) |\xi\rangle\langle\bar{\xi}|$$

Here the integration is taken over the entire complex plane with the measure $d\xi_n d\bar{\xi}_n = dxdy$ *for* $\xi_n = x_n + iy_n$.

(5) *For arbitrary invertible constant* 2×2 *matrices* $\mathscr{A}_n$ *and two components constant column vectors* $\mathscr{B}_n$*, one has*

$$\int \prod_{n>0} \frac{d\xi_n\bar{\xi}_n}{\mathscr{C}_n} \exp\left(-\frac{1}{2}\sum_{n>0} \frac{1}{\mathscr{C}_n} (\bar{\xi}_n\ \xi_n)\, \mathscr{A}_n \begin{pmatrix} \bar{\xi}_n \\ \xi_n \end{pmatrix} + \sum_{n>0} (\bar{\xi}_n\ \xi_n)\, \mathscr{B}_n\right)$$

$$= \prod_{n>0} (-\det \mathscr{A}_n)^{-1/2} \exp\left(\frac{1}{2}\sum_{n>0} \mathscr{C}_n\, {}^t\mathscr{B}_n \mathscr{A}_n^{-1} \mathscr{B}_n\right)$$

E.2 Trace Formula

Proposition E.2.1

$$\mathrm{tr}_{\mathscr{F}} q^{\kappa\bar{d}} \exp\left(\sum_{m>0} \alpha_{-m}\zeta_{-m}\right) \exp\left(\sum_{m>0} \alpha_m\zeta_m\right).$$

$$= \frac{1}{(q^\kappa; q^\kappa)_\infty} \exp\left(\sum_{n>0} \frac{\mathscr{C}_n q^{\kappa n}}{1 - q^{\kappa n}} \zeta_{-n}\zeta_n\right) \tag{E.2.1}$$

Proof Noting

$$\mathrm{tr}_{\mathscr{F}} \mathscr{O} = \int \prod_{n>0} \frac{d\xi_n \bar{\xi}_n}{\mathscr{C}_n} \exp\left(-\sum_{n>0} \frac{1}{\mathscr{C}_n} \bar{\xi}_n \xi_n \right) \langle \bar{\xi} | \mathscr{O} | \xi \rangle$$

and (3) and use (5) for

$$\mathscr{A}_n = \begin{pmatrix} 0 & 1 - q^{\kappa n} \\ 1 - q^{\kappa n} & 0 \end{pmatrix}, \qquad \mathscr{B}_n = \begin{pmatrix} q^{\kappa n} \zeta_{-n} \\ \zeta_n \end{pmatrix}. \qquad \square$$

References

1. T.-D. Albert, H.Boos, R.Flume, R. Poghossian, K. Ruhlig, An F-twisted XYZ model. Lett. Math. Phys. **53**, 201–214 (2000)
2. L.F. Alday, D. Gaiotto, Y. Tachikawa, Liouville correlation functions from four dimensional gauge theories, Lett. Math. Phys. **91**, 167–197 (2010)
3. M. Aganagic, A. Okounkov, Elliptic Stable Envelopes. Preprint (2016). arXiv:1604.00423
4. M. Aganagic, A. Okounkov, Quasimap counts and Bethe eigenfunctions. Preprint (2017). arXiv:1704.08746
5. K. Aomoto, M. Kita, *Theory of Hypergeometric Functions*, 1994, Maruzen Pub. (in Japanese); English Edition, 2011. *Springer Monographs in Mathematics* (Springer, Japan, 1994)
6. Y. Asai, M. Jimbo, T. Miwa, Y. Pugai, Bosonization of vertex operators for the $A^{(1)}_{n-1}$ face model. J. Phys. A **29**, 6595–6616 (1996)
7. G.E. Andrews, R.J. Baxter, P.J. Forrester, Eight vertex SOS model and generalized Rogers-Ramanujan-type identities. J. Stat. Phys. **35**, 193–266 (1984)
8. D. Arnaudon, E. Buffenoir, E. Ragoucy, Ph. Roche, Universal solutions of quantum dynamical Yang-Baxter equations. Lett. Math. Phys. **44**, 201–214 (1998)
9. H. Awata, H. Kubo, S. Odake, J. Shiraishi, Quantum W_N algebras and Macdonald polynomials. Comm. Math. Phys. **179**, 401–416 (1996)
10. O. Babelon, D. Bernard, E. Billey, A quasi-Hopf algebra interpretation of quantum $3j$- and $6j$-symbols and difference equations. Phys. Lett. B **375**, 89–97 (1996)
11. J. Bagger, D. Nemeschansky, S. Yankielowicz, Virasoro algebras with central charge $c > 1$. Phys. Rev. Lett. **60**, 389–392 (1988)
12. R.J. Baxter, *Exactly Solved Models in Statistical Mechanics* (Academic Press, London, 1982)
13. A.A. Belavin, Dynamical symmetry of integrable quantum systems. Nucl. Phys. **B180**[FS2], 189–200 (1981)
14. A.A. Belavin, V.G. Drinfeld, Triangle equations and simple Lie algebras, in *Classic Reviews in Mathematics and Mathematical Physics*, vol. 1 (Harwood Academic Publishers, Amsterdam, 1998)
15. A.A. Belavin, A.M. Polyakov, A.B. Zamolodchikov, Infinite conformal symmetry in two-dimensional quantum field theory. *Nucl. Phys. B* **241**, 333–380 (1984)
16. A. Bertram, I. Ciocan-Fontainine, B. Kim, Gromov-Witten invariants for abelian and nonabelian quotients. J. Algebraic Geom. **17**, 275–294 (2008)
17. A. Bougourzi, L. Vinet, A quantum analog of the $\mathscr{Z}$ algebra. J. Math. Phys. **37**, 3548–3567 (1996)
18. P. Bouwknegt, K. Schoutens, W symmetry in conformal field theory. Phys. Rep. **223**, 183–276 (1993)

H. Konno, *Elliptic Quantum Groups*, SpringerBriefs in Mathematical Physics 37,
https://doi.org/10.1007/978-981-15-7387-3

19. J.-S. Caux, H. Konno, M. Sorrell, R. Weston, Tracking the effects of interactions on spinons in gapless Heisenberg chains. Phys. Rev. Lett. **106**, 217203 (4 p.) (2011)
20. J.-S. Caux, H. Konno, M. Sorrell, R. Weston, Exact form-factor results for the longitudinal structure factor of the massless XXZ model in zero field. J. Stat. Mech., P01007 (40 p.) (2012)
21. V. Chari, A. Pressley, *A Guide to Quantum Groups* (Cambridge Univ. Press., Cambridge, 1994)
22. V. Chari, A. Pressley, Yangians and R-matrices. *L'Enseignement Math.* **36**, 267–302 (1990)
23. V. Chari, A. Pressley, Quantum affine algebras. Comm. Math. Phys. **142**, 261–283 (1991)
24. N. Chriss, V. Ginzburg, *Representation Theory and Complex Geometry* (Birkhäuser, 1994)
25. E. Date, M. Jimbo, T. Miwa, M. Okado, Fusion of the eight vertex SOS model. Lett. Math. Phys. **12**, 209–215 (1986)
26. E. Date, M. Jimbo, A. Kuniba, T. Miwa, M. Okado, Exactly solvable SOS models. Nucl. Phys. B **290**[FS20], 231–273 (1987); Exactly solvable SOS models II. Adv. Stud. Pure Math. **16**, 17–122 (1988)
27. P. Difrancesco, H. Saleur, J.-B. Zuber, Generalized Coulomb-Gas formalism for two dimensional critical models based on $SU(2)$ coset construction. Nucl. Phys. **B300**, 393–432 (1988)
28. J. Ding, I.B. Frenkel, Isomorphism of two realizations of quantum affine algebra $U_q(\widehat{\mathfrak{gl}(n)})$. Comm. Math. Phys. **156**, 277–300 (1993)
29. V.G. Drinfeld, Quantum groups. Proc. ICM Berkeley **1**, 789–820 (1986)
30. V.G. Drinfeld, A new realization of Yangians and quantized affine algebras. Sov. Math. Dokl. **36**, 212–216 (1988)
31. V.G. Drinfeld, Quasi-Hopf algebras, Leningrad Math. J. **1**, 1419–1457 (1990)
32. V.G. Drinfeld, On quasitriangular quasi-Hopf algebras and a group closely connected with Gal($\overline{\mathbf{Q}}/\mathbf{Q}$). Leningrad Math. J. **2**, 829–860 (1991)
33. B. Enriquez, G. Felder, Elliptic quantum groups $E_{\tau,\eta}(\mathfrak{sl}_2)$ and quasi-Hopf algebra. Comm. Math. Phys. **195**, 651–689 (1998)
34. B. Enriquez, V.N. Rubtsov, Quantum groups in higher genus and Drinfeld's new realizations method (sl2 case). Ann. Sci. École Norm. Sup. **30**, 821–846 (1997); Quasi-Hopf algebras associated with sl_2 and complex curves. Isr. J. Math. **112**, 61–108 (1999)
35. P. Etingof, I. Frenkel, A. Kirillov, Jr., Lectures on representation theory and Knizhnik-Zamolodchikov equations. *Mathematical Surveys and Monographs*, vol. 58 (AMS, Providence, Rhode Island, 1998)
36. P. Etingof, A. Varchenko, Solutions of the quantum dynamical Yang-Baxter equation and dynamical quantum groups. Comm. Math. Phys. **196**, 591–640 (1998); Exchange dynamical quantum groups. Comm. Math. Phys. **205**, 19–52 (1999)
37. L.D. Faddeev, N.Yu. Reshetikhin, L.A. Takhtadjan, Quantization of Lie groups and Lie algebras. Leningrad Math. J. **1**, 178–201 (1989)
38. R.M. Farghly, H. Konno, K. Oshima, Elliptic algebra $U_{q,p}(\widehat{\mathfrak{g}})$ and quantum Z-algebras. Algebr. Represent. Theory **18**, 103–135 (2014)
39. B. Feigin, M. Finkelberg, I. Frenkel, L. Rybnikov, Gelfand-Tsetlin algebras and cohomology rings of Laumon spaces. Sel. Math. New Ser. **17**, 337–361 (2011)
40. B. Feigin, M. Finkelberg, A. Negut, L. Rybnikov, Yangians and cohomology rings of Laumon spaces. Sel. Math. New Ser. **17**, 573–607 (2011)
41. B. Feigin, E. Frenkel, Quantum W-algebras and elliptic algebras. Comm. Math. Phys. **178**, 653–678 (1996)
42. B. Feigin, D. Fuchs, Skew-symmetric invariant differential operators on the line and Verma modules over the Virasoro algebra. Funktsional. Anal. i Prilozhen. **16**, 47–63 (1982). English translation: Funct. Annal. Appl. **16**, 114–126 (1982); Verma modules over a Virasoro algebra. Funktsional. Anal. i Prilozhen. **17**, 91–92 (1983). English translation: Funct. Annal. Appl. **17**, 241–242 (1983)

43. B. Feigin, M. Jimbo, T. Miwa, A. Odesskii, Y. Pugai, Algebra of screening operators for the deformed W_n algebra. Comm. Math. Phys. **191**, 501–541 (1998)
44. B. Feigin, A. Tsymbaliuk, Bethe subalgebras of $U_q(\widehat{\mathfrak{gl}}_n)$ via shuffle algebras. Sel. Math. (N.S.) **22**(2), 979–1011 (2016)
45. G. Felder, Elliptic quantum groups. Proc. ICMP Paris-1994, 211–218 (1995)
46. G. Felder, Conformal field theory and integrable systems associated to elliptic curves, in *Proc. ICM Zürich -1994*, pp. 1247–1255 (Birkhäuser, Basel, 1995)
47. G. Felder, A. Varchenko, On representations of the elliptic quantum group $E_{\tau,\eta}(sl_2)$. Comm. Math. Phys. **181**, 741–761 (1996)
48. G. Felder, A. Varchenko, Algebraic Bethe ansatz for the elliptic quantum group $E_{\tau,\eta}(sl_2)$. Nucl. Phys. **B 480**, 485–503 (1996)
49. G. Felder, A. Varchenko, Elliptic quantum groups and Ruijsenaars models. J. Stat. Phys. **89**, 963–980 (1997)
50. G. Felder, V. Tarasov, A. Varchenko, Solutions of the elliptic QKZB equations and Bethe ansatz I, in *Topics in Singularity Theory, V.I.Arnold's 60th Anniversary Collection*. Advances in the Mathematical Sciences. AMS Translations, Series 2, vol. 180, pp. 45–76 (1997)
51. G. Felder, R. Rimányi, A. Varchenko, Elliptic Dynamical Quantum Groups and Equivariant Elliptic Cohomology. Preprint (2017). arXiv:1702.08060
52. O. Foda, K. Iohara, M. Jimbo, R. Kedem, T. Miwa, H. Yan, An elliptic quantum algebra for sl_2. Lett. Math. Phys. **32**, 259–268 (1994)
53. O. Foda, K. Iohara, M. Jimbo, R. Kedem, T. Miwa, H. Yan, Notes on highest weight modules of the elliptic algebra $\mathscr{A}_{q,p}(\widehat{sl}_2)$. Quantum field theory, integrable models and beyond (Kyoto, 1994). Progr. Theoret. Phys. Suppl. **118**, 1–34 (1995)
54. O. Foda, M. Jimbo, T. Miwa, K. Miki, A. Nakayashiki, Vertex operators in solvable lattice models. J. Math. Phys. **35**, 13–46 (1994)
55. I.B. Frenkel, N.H. Jing, Vertex representations of quantum affine algebras. Proc. Natl. Acad. Sci. USA, **85**, 9373–9377 (1988)
56. I.B. Frenkel, N. Reshetikhin, Quantum affine algebras and holonomic difference equations. Comm. Math. Phys. **146**, 1–60 (1992)
57. E. Frenkel, N. Reshetikhin, Deformation of W-Algebras Associated to Simple Lie Algebras. arXiv:q-alg/9708006
58. C. Frønsdal, Generalization and exact deformations of quantum groups. Publ. RIMS Kyoto Univ. **33**, 91–149 (1997)
59. C. Frønsdal, Quasi-Hopf deformations of quantum groups. Lett. Math. Phys. **40**, 117–134 (1997)
60. N. Ganter, The elliptic Weyl character formula. Compos. Math. **150**, 1196–1234 (2014)
61. S. Gautam, V. Toledano Laredo, Elliptic Quantum Groups and Their Finite-Dimensional Representations. Preprint (2019). arXiv:1707.06469v2
62. J.L. Gervais, A. Neveu, Novel triangle relation and absence of tachyons in Liouville field theory. Nucl. Phys. **B238**, 125–141 (1984)
63. V. Ginzburg, Lagrangian construction of the enveloping algebra $U(sl_n)$. C. R. Acad. Sci. **312**, 907–912 (1991)
64. V. Ginzburg, E. Vasserot, Langlands reciprocity for affine quantum groups of type A_n. Int. Math. Res. Notices **3**, 67–85 (1993)
65. V. Ginzburg, M. Kapranov, E. Vasserot, Elliptic Algebras and Equivariant Elliptic Cohomology I. Preprint (1995). arXiv:q-alg/9505012
66. A. Givental, B. Kim, Quantum cohomology of flag manifolds and Toda lattices. Comm. Math. Phys. **168**, 609–641 (1995)
67. A. Givental, Y-P.Lee, Quantum K-theory on flag manifolds, finite-difference Toda lattices and quantum groups. Invent. Math. **151**, 193–219 (2003)

68. P. Goddard, A. Kent, D. Olive, Virasoro algebras and coset space models. Phys. Lett. B **152**, 88–92 (1985); Unitary representations of the Virasoro and super-Virasoro algebras. Comm. Math. Phys. **103**, 105–119 (1986)
69. V. Gorbounov, R. Rimányi, V. Tarasov, A. Varchenko, Cohomology of the cotangent bundle of a flag variety as a Yangian Bethe algebra. J. Geom. Phys. **74**, 56–86 (2013)
70. I. Grojnowski, Delocalised equivariant elliptic cohomology, in *Elliptic Cohomology*, vol. 342 of *London Math. Soc. Lecture Note Ser.*, pp. 114–121 (2007)
71. J. Hartwig, The elliptic $GL(n)$ dynamical quantum group as an $\mathfrak{h}$-Hopf algebroid. Int. J. Math. Math. Sci. **Art. ID 545892**, 41 pp. (2009)
72. M. Idzumi, K. Iohara, M. Jimbo, T. Miwa, A. Nakayashiki, T. Tokihiro, Quantum affine symmetry in vertex models. Int. J. Mod. Phys. **A8**, 1479–1511 (1993)
73. M. Jimbo, A q-difference analogue of $U_q(\mathfrak{g})$ and the Yang-Baxter equation. Lett. Math. Phys. **10**, 63–69 (1985)
74. M. Jimbo, A. Kuniba, T. Miwa, M. Okado, The $A_n^{(1)}$ face models. Comm. Math. Phys. **119**, 543–565 (1988)
75. M. Jimbo, H. Konno, T. Miwa, Massless XXZ model and degeneration of the elliptic algebra $\mathscr{A}_{q,p}(\widehat{\mathfrak{sl}}_2)$, in *Deformation Theory and Symplectic Geometry (Ascona, 1996)*. Math. Phys. Stud., vol. 20 (Kluwer Acad. Publ., Dordrecht, 1997), pp. 117–138
76. M. Jimbo, H. Konno, S. Odake, J. Shiraishi, Elliptic algebra $U_{q,p}(\widehat{\mathfrak{sl}}_2)$: Drinfeld currents and vertex operators. Comm. Math. Phys. **199**, 605–647 (1999)
77. M. Jimbo, H. Konno, S. Odake, J. Shiraishi, Quasi-Hopf twistors for elliptic quantum groups. Transformation Groups **4**, 303–327 (1999)
78. M. Jimbo, H. Konno, S. Odake, Y. Pugai, J. Shiraishi, Free field construction for the ABF models in regime II. J. Stat. Phys. **102**, 883–921 (2001)
79. M. Jimbo, T. Miwa, *Algebraic Analysis of Solvable Lattice Models*. CBMS Regional Conference Series in Mathematics, vol. 85 (AMS, 1994)
80. M. Jimbo, T. Miwa, A. Nakayashiki, Difference equations for the correlation functions of the eight-vertex model. J. Phys. A **26**, 2199–2209 (1993)
81. M. Jimbo, T. Miwa, M. Okado, Solvable lattice models related to the vector representation of classical simple Lie algebras. Comm. Math. Phys. **116** 507–525 (1988)
82. N. Jing, Higher level representations of the quantum affine algebra $U_q(\widehat{sl}(2))$. J. Algebra **182**, 448–468 (1996); Quantum z-algebras and representations of quantum affine algebras. Comm. Alg. **28**, 829–844 (2000)
83. V.G. Kac, *Infinite Dimensional Lie Algebras*, 3rd edn. (Cambridge University Press, Cambridge, 1990)
84. D. Kastor, E. Martinec, Z. Qiu, Current algebra and conformal discrete series. Phys. Lett. B **200**, 434–440 (1988)
85. A. Kirillov Jr., Quiver representations and quiver varieties, in *Graduate Studies in Math.*, vol. 174 (AMS, Providence, RI, 2016)
86. E. Koelink, H. Rosengren, Harmonic analysis on the $SU(2)$ dynamical quantum group. Acta. Appl. Math. **69**, 163–220 (2001)
87. E. Koelink, Y.van Norden, H. Rosengren, Elliptic $U(2)$ quantum group and elliptic hypergeometric series. Comm. Math. Phys. **245**, 519–537 (2004)
88. E. Koelink, Y. van Norden, Pairings and actions for dynamical quantum group. Adv. Math. **208**, 1–39 (2007)
89. R. Kodera, Affine Yangian action on the Fock Space. Preprint (2015). arXiv:1506.01246
90. T. Kojima, H. Konno, The elliptic algebra $U_{q,p}(\widehat{\mathfrak{sl}}_N)$ and the Drinfeld realization of the elliptic quantum group $\mathscr{B}_{q,\lambda}(\widehat{\mathfrak{sl}}_N)$. Comm. Math. Phys. **239**, 405–447 (2003)
91. T. Kojima, H. Konno, The Drinfeld realization of the elliptic quantum group $\mathscr{B}_{q,\lambda}(A_2^{(2)})$. J. Math. Phys. **45**, 3146–3179 (2004)
92. T. Kojima, H. Konno, The elliptic algebra $U_{q,p}(\widehat{\mathfrak{sl}}_2)$ and the deformation of W_N algebra. J. Phys. A **37**, 371–383 (2004)

93. T. Kojima, H. Konno, R. Weston, The vertex-face correspondence and correlation functions of the fusion eight-vertex models I: The general formalism. Nucl. Phys. **B720**, 348–398 (2005)
94. H. Konno, An elliptic algebra $U_{q,p}(\widehat{\mathfrak{sl}}_2)$ and the fusion RSOS models. Comm. Math. Phys. **195**, 373–403 (1998)
95. H. Konno, Dynamical R matrices of elliptic quantum groups and connection matrices for the q-KZ equations. SIGMA, **2**, Paper 091, 25 pages (2006)
96. H. Konno, Elliptic quantum group $U_{q,p}(\widehat{\mathfrak{sl}}_2)$ and vertex operators. J. Phys. A **41**, 194012 (2008)
97. H. Konno, Elliptic quantum group $U_{q,p}(\widehat{\mathfrak{sl}}_2)$, Hopf algebroid structure and elliptic hypergeometric series. J. Geom. Phys. **59**, 1485–1511 (2009)
98. H. Konno, Elliptic quantum group, Drinfeld coproduct and deformed W-algebras, in *Recent Advances in Quantum Integrable Systems 2014*, Dijon
99. H. Konno, Elliptic weight functions and elliptic q-KZ equation. J. Integrable Syst. **2**, 1–43 (2017). https://doi.org/10.1093/integr/xyx011
100. H. Konno, Elliptic stable envelopes and finite-dimensional representations of elliptic quantum group. J. Integrable Syst. **3**, 1–43 (2018). https://doi.org/10.1093/integr/xyy012
101. H. Konno, Elliptic quantum groups $U_{q,p}(\widehat{\mathfrak{gl}}_N)$ and $E_{q,p}(\widehat{\mathfrak{gl}}_N)$. Adv. Stud. Pure Math. **76**, 347–417 (2018)
102. H. Konno, Talks given at the workshops, in *Elliptic Cohomology Days*, June 10–15, 2019, Univ. of Illinois, Urbana–Champaign; *Elliptic Integrable Systems, Special Functions and Quantum Field Theory*, June 16–20, 2019, Nordita, Stockholm; *Representation Theory and Integrable Systems*, August 12–16, 2019, ETH, Zürich
103. H. Konno, K. Oshima, Elliptic quantum group $U_{q,p}(B_N^{(1)})$ and vertex operators. RIMS Kokyuroku Bessatsu **B62**, 97–148 (2017)
104. H. Konno, K. Oshima, Elliptic Quantum Toroidal Algebras, in preparation
105. P. Koroteev, P. Pushkar, A. Smirnov, A. Zeitlin, Quantum K-theory of Quiver Varieties and Many-body Systems. Preprint (2017). arXiv:1705.10419
106. A. Kuniba, Exact solution of solid-on-solid models for twisted affine Lie algebras $A_{2n}^{(2)}$ and $A_{2n-1}^{(2)}$. Nucl. Phys. **B355**, 801–821 (1991)
107. A. Kuniba, J. Suzuki, Exactly solvable $G_2^{(1)}$ solid-on-solid models. Phys. Lett. **A160**, 216–222 (1991)
108. H. Lange, Ch. Birkenhake, *Complex Abelian Varieties* (Springer, 1992)
109. M. Lashkevich, Y. Pugai, Free field construction for correlation functions of the eight-vertex model. Nucl. Phys. **B516**, 623–651 (1998)
110. J. Lepowsky, R.L. Wilson, A new family of algebras underlying the Rogers-Ramanujan identities and generalizations. Proc. Natl. Acad. Sci. USA **78**, 7254–7258 (1981); The structure of standard modules, I: Universal algebras and the Roger-Ramanujan identities. Invent. Math. **77**, 199–290 (1984)
111. S. Lukyanov, Free field representation for massive integrable models. Comm. Math. Phys. **167**, 183–226 (1995)
112. S.L. Lukyanov, V.A. Fateev, Additional symmetries and exactly-soluble models in two-dimensional conformal field theory. Sov. Sci. Rev. A. Phys. **15**, 1–117 (1990)
113. S. Lukyanov, Y. Pugai, Multi-point local height probabilities in the integrable RSOS model. Nucl. Phys. **B473**, 631–658 (1996)
114. J.-M. Maillet, J. Sanchez de Santos, Drinfeld twists and algebraic Bethe ansatz. Am. Math. Soc. Transl. **201**, 137–178 (2000)
115. A. Matsuo, Jackson integrals of Jordan Pochhammer type and quantum Knizhnik Zamolodchikov equations. Comm. Math. Phys. **151**, 263–273 (1993); Quantum algebra structure of certain Jackson integrals. Comm. Math. Phys. **157**, 479–498 (1993)
116. A. Matsuo, A q-deformation of Wakimoto modules, primary fields and screening operators. Comm. Math. Phys. **160**, 33–48 (1994)
117. D. Maulik, A. Okounkov, Quantum Groups and Quantum Cohomology. Preprint (2012). arXiv:1211.1287

118. K. Mcgerty, T. Nevins, Kirwan Surjectivity for Quiver Varieties. Preprint (2016). arXiv:1610.08121
119. K. Mimachi, A solution to quantum Knizhnik-Zamolodchikov equations and its application to eigenvalue problems of the Macdonald type. Duke Math. J. **85**, 635–658 (1996)
120. K. Mimachi, M. Noumi, Representations of the Hecke Algebra on a Family of Rational Functions. Preprint (1996). Unpublished
121. T. Miwa, R. Weston, Boundary ABF models. Nucl. Phys. B **486**, 517–545 (1997)
122. A. Molev, in Yangians and Classical Lie algebras. Mathematical Surveys and Monographs, vol. 143 (AMS, 2007)
123. K. Nagao, K-theory of Quiver varieties, q-Fock space and nonsymmetric Macdonald polynomials. Osaka J. Math. **46**, 877–907 (2009)
124. H. Nakajima, Instantons on ALE spaces, Quiver varieties and Kac-Moody algebras. Duke Math. J. **76**, 365–416 (1994)
125. H. Nakajima, Quiver varieties and Kac-Moody algebras. Duke Math. J. **91**, 515–560 (1998)
126. H. Nakajima, Quiver varieties and finite-dimensional representations of quantum affine algebras. J. Am. Math. Soc. **14**, 145–238 (2001)
127. H. Nakajima, Lectures on Perverse Sheaves on Instanton Moduli Spaces. Preprint (2016). arXiv:1604.06316
128. A. Negut, The shuffle algebra revisited. Int. Math. Res. Not. **22**, 6242–6275 (2014); Quantum toroidal and shuffle algebras, R-matrices and a conjecture of Kuznetsov. Preprint (2013). arXiv:1302.6202
129. N. Nekrasov, S. Shatashivili, Bethe ansatz and supersymmetric vacua. AIP Conf. Proc. **1134**, 154–169 (2009); Quantum integrability and supersymmetric vacua, Prog. Theor. Phys. Suppl. **177**, 105–119 (2009): Supersymmetric vacua and Bethe ansatz Nucl. Phys. Proc. Suppl. **192–193**, 91–112 (2009)
130. A. Okounkov, Lectures on K-theoretic Computations in Enumerative Geometry (2015). arXiv: 1512.07363
131. S. Pakuliak, V. Rubtsov, A. Silantyev, The SOS model partition function and the elliptic weight functions. J. Phys. A **41**, 295204, 20 pp. (2008)
132. F. Ravanini, An infinite class of new conformal field theories with extended algebras Mod. Phys. Lett. A **3**, 397–412 (1988)
133. N.Yu. Reshetikhin, M.A. Semenov-Tian-Shansky, Central extensions of quantum current groups. Lett. Math. Phys. **19**, 133–142 (1990)
134. R. Rimányi, A. Smirnov, A. Varchenko, Z. Zhou, Three-dimensional mirror self-symmetry of the cotangent bundle of the full flag variety. SIGMA Symmetry Integrability Geom. Methods Appl. **15**, 093, 22 p. (2019)
135. R. Rimányi, V. Tarasov, A. Varchenko, Trigonometric weight functions as K-theoretic stable envelope maps for the cotangent bundle of a flag variety. J. Geom. Phys. **94**, 81–119 (2015)
136. R. Rimányi, A. Varchenko, Dynamical Gelfand-Tetlin Algebra and Equivariant Cohomology of Grassmannians. Preprint (2015). arXiv:1510.03625
137. R. Rimányi, V. Tarasov, A. Varchenko, Elliptic and K-theoretic Stable Envelopes and Newton Polytopes. Preprint (2017). arXiv:1705.09344
138. H. Rosengren, Sklyanin invariant integration. Int. Math. Res. Not. **60**, 3207–3232 (2004)
139. H. Rosengren, Felder's elliptic quantum group and elliptic hypergeometric series on the root system A_n. Int. Math. Res. Not. **13**, 2861–2920 (2011)
140. H. Rosengren, An Izergin-Korepin-type identity for the 8VSOS model, with applications to alternating sign matrices. Adv. Appl. Math. **43**, 137–155 (2009)
141. I. Rosu, Equivariant elliptic cohomology and rigidity. Am. J. Math. **123**, 647–677 (2001)
142. V. Rubtsov, A. Silantyev, D. Talalaev, Manin matrices, quantum elliptic commutative families and characteristic polynomial of elliptic Gaudin model. SIGMA Symmetry Integrability Geom. Methods Appl. **5**, Paper 110, 22 pp. (2009)
143. S.N.M. Ruijsenaars, First order analytic difference equations and integrable quantum systems. J. Math. Phys. **38**, 1069–1146 (1997)

144. E. Sklyanin, Some algebraic structures connected with the Yang-Baxter equation. Funct. Anal. Appl. **16**, 263–270 (1983)
145. E.K. Sklyanin, Some algebraic structures connected with the Yang-Baxter equation. Representations of quantum algebras. Funct. Anal. Appl. **17**, 273–284 (1984)
146. D. Shenfeld, Abelianization of Stable Envelopes in Symplectic Resolutions, Ph.D. Thesis, Princeton University, 2013
147. J. Shiraishi, H. Kubo, H. Awata, S. Odake, A quantum deformation of the Virasoro algebra and the Macdonald symmetric functions. Lett. Math. Phys. **38**, 33–51 (1996)
148. A. Smirnov, Polynomials associated with fixed points on the instanton moduli space. Preprint (2014). arXiv:1404.5304
149. A. Smirnov, Elliptic stable envelope for Hilbert scheme of points in the plane. *Sel. Math. (N.S.)* **26**(1), Paper No. 3 (2020)
150. F. Smirnov, Form factors in completely integrable models of quantum field theory, in *Advanced Series in Mathematical Physics*, vol. 14 (World Scientific, 1992)
151. T. Takebe, Q-operators for higher spin eight vertex models with a rational anisotropy parameter. Lett. Math. Phys. **109**, 186–1890 (2019)
152. T. Tanisaki, Killing forms, Harish-Chandra isomorphisms, and universal R-matrices for quantum algebras. Int. J. Mod. Phys. **A7**(supplement 1B), 941–961 (1992)
153. V. Tarasov, A. Varchenko, Geometry of q-hypergeometric functions, quantum affine algebras and elliptic quantum groups. Astérisque **246** (1997). Société Mathématique de France
154. A. Tsymbaliuk, Quantum affine Gelfand-Tsetlin bases and quantum toroidal algebra via K-theory of affine Laumon Spaces. Sel. Math. New. Ser. **16**, 173–200 (2010)
155. D. Uglov, Symmetric functions and the Yangian decomposition of the Fock and basic modules of the affine Lie algebra $\widehat{\mathfrak{sl}}_N$, quantum many-body problems and representation theory. MSJ Memoirs Math. Soc. Jpn. Tokyo **1**, 183–241 (1998)
156. Y. van Norden, Dynamical Quantum Groups, Duality and Special Functions, Ph.D. Thesis, 2005
157. M. Varagnolo, E. Vasserot, On the K-theory of the cyclic Quiver variety. Int. Math. Res. Notices **18**, 1005–1028 (1999)
158. M. Varagnolo, Quiver varieties and Yangians. Lett. Math. Phys. **53**, 273–283 (2000)
159. E. Vasserot, Affine quantum groups and equivariant K-theory. Transform. Groups **3**, 269–299 (1998)
160. A.B. Zamolodchikov, V.A. Fateev, Representations of the algebra of "parafermion currents" of spin 4/3 in two-dimensional conformal field theory. Minimal models and the tricritical potts $\mathbb{Z}_3$ model. Theor. Math. Phys. **71**, 451–462 (1987)

Printed in the United States
By Bookmasters